Nanomedicine in Diagnostics

Nanomedicine in Diagnostics

Editor
Noemi Rozlosnik
Associate Professor
Department of Micro- and Nanotechnology
Technical University of Denmark
Kongens Lyngby
Denmark

CRC Press
Taylor & Francis Group
an informa business
www.taylorandfrancisgroup.com

6000 Broken Sound Parkway, NW
Suite 300, Boca Raton, FL 33487
711 Third Avenue
New York, NY 10017
2 Park Square, Milton Park
Abingdon, Oxon OX14 4RN, UK

Science Publishers
Jersey, British Isles
Enfield, New Hampshire

Published by Science Publishers, an imprint of Edenbridge Ltd.

- St. Helier, Jersey, British Channel Islands
- P.O. Box 699, Enfield, NH 03748, USA

E-mail: *info@scipub.net* Website: *www.scipub.net*

Marketed and distributed by:

CRC Press
Taylor & Francis Group
an **informa** business
www.taylorandfrancisgroup.com

6000 Broken Sound Parkway, NW
Suite 300, Boca Raton, FL 33487
711 Third Avenue
New York, NY 10017
2 Park Square, Milton Park
Abingdon, Oxon OX14 4RN, UK

ISBN 978-1-57808-738-9

Cover Illustrations

The Publisher and the Editor express their thanks to the following persons for use of their figures: Xueli Liu, Li Chen and Shutao Wang, Jaime Castillo-León, Yuri Belotti and Winnie E. Svendsen, Karsten Damstedt, Jesper Scheel, Pravin K. Naoghare and Joon Myong Song

Library of Congress Cataloging-in-Publication Data

Nanomedicine in diagnostics / editor, Noemi Rozlosnik.
p. ; cm.
Includes bibliographical references and index.
ISBN 978-1-57808-738-9 (hardcover : alk. paper)
I. Rozlosnik, Noemi.
[DNLM: 1. Nanostructures--diagnostic use. 2. Nanomedicine--methods. QT 36.5]
610.28--dc23

2011038909

Printed in the United States of America

Preface

Nanomedicine is a young and rapidly evolving field of research at the cross-roads of nanoscience and medicine, and has tremendous potential in disease diagnosis, imaging, monitoring, and therapy. The possibility of exploiting nanotechnological tools for novel functional materials, biosensors, and cell based detection systems has created a growing field in medical diagnostics.

The new concepts of nanodimensions have opened the door to a wide variety of microsystems for clinical applications. Such devices are extremely useful for delivering diagnostic information in a fast, simple, and low cost manner. Current research is exploring the design and fabrication of functional nanostructures, nanomotors, sensors at molecular scale and microstructured electrochemical systems.

This book provides the first collection on topics in nanomedicine, where diagnostics is the main goal. All individual chapters were written by leading world experts in corresponding fields. The editor is extremely grateful to all the contributors for providing the state-of-the-art manuscript.

The primary intended readers of this book are biomedical engineers, researchers, clinical laboratory analysts, and other technical and professional people, who are interested in the future of medical technology.

Designed as a broad survey of the field, this book combines contributions in a wide range of studies from single molecule diagnostics up to cell based systems. The focus of the first three chapters is the functionality of particular nanosized structures. The next few chapters guide the reader in the new trends of biosensors, while the last part shows examples on cell based diagnostic systems. Although this book is not able to cover all the topics currently being developed and under development completely, but we believe, it does provide a good representation of the field. We hope it will facilitate the collaboration of scientists, engineers and clinical researchers for further development of nanomedicine in diagnostics.

In our opinion, nanomedicine has reached a new level of maturity: in the near future the integration of diagnostic approaches into the framework of nanotechnology will revolutionize medical diagnostics.

Copenhagen, July 2011 — Noemi Rozlosnik

Contents

List of Contributors

A.I. Archakov
Institute of Biomedical Chemistry, Russian Academy of Medical Sciences, Pogodinskaya St. 10, Moscow 119121, Russia.
E-mail: Archakov@ibmc.msk.ru

Yuri Belotti
Department of Micro and Nanotechnology, Technical University of Denmark, Building 345 east, Lyngby 2800, Denmark.
E-mail: s102136@student.dtu.dk

T.V. Bulko
Institute of Biomedical Chemistry, Russian Academy of Medical Sciences, Pogodinskaya St. 10, Moscow 119121, Russia.
E-mail: tanabulko@mail.ru

Jaime Castillo-León
Department of Micro and Nanotechnology, Technical University of Denmark, Building 345 east, Lyngby 2800, Denmark.
E-mail: jaic@nanotech.dtu.dk

Gianfranco Cerofolini
Department of Materials Science, University of Milano Bicocca, Via R. Cozzi 53, 20125 Milano, Italy.
E-mail: gianfranco.cerofolini@mater.unimib.it

Ya. M. Chalenko
Institute of Biomedical Chemistry, Russian Academy of Medical Sciences, Pogodinskaya St. 10, Moscow 119121, Russia.
E-mail: yaroslavazaka@yandex.ru

Li Chen
Beijing National Laboratory for Molecular Sciences (BNLMS), Institute of Chemistry, Chinese Academy of Sciences, Beijing 100190 P.R. China.
E-mail: Lichen@iccas.ac.cn

Nikolaj Ormstrup Christiansen
Department of Micro-and Nanotechnology, Technical University of Denmark, Produktionstorvet 423, DK-2800 Lyngby Denmark.
E-mail: Nikolaj.christiansen@nanotech.dtu.dk

Johannes Daprà
Department of Micro-and Nanotechnology, Technical University of Denmark, Produktionstorvet 423, DK-2800 Lyngby Denmark.
E-mail: jdap@nanotech.dtu.dk

Li-Qun Gu
Department of Biological Engineering and Dalton Cardiovascular Research Center, University of Missouri, Columbia, MO 65201.
E-mail: gul@missouri.edu

Patrizia Iavicoli
Institut for Bioengineering of Catalonia (IBEC). C/Baldiri Reixac 10-12, 08028 Barcelona, Spain.
E-mail: piavicoli@ibecbarcelona.eu

Akira Ishihama
Hosei University, Department of Frontier Bioscience, Koganei, Tokyo 184-8584, Japan.
E-mail: aishiham@hosei.ac.jp

Katrine Kiilerich-Pedersen
Department of Micro-and Nanotechnology, Technical University of Denmark, Produktionstorvet 423, DK-2800 Lyngby Denmark.
E-mail: katk@nanotech.dtu.dk

Xueli Liu
Beijing National Laboratory for Molecular Sciences (BNLMS), Institute of Chemistry, Chinese Academy of Sciences, Beijing 100190 (P.R. China).
E-mail: liuxl@iccas.ac.cn

Pravin K. Naoghare
Research Institute of Pharmaceutical Sciences and College of Pharmacy, Seoul National University, Seoul 151-742, South Korea.
E-mail: pk_naoghare@neeri.res.in; pinbiotech@gmail.com

Dario Narducci
Department of Materials Science, University of Milano Bicocca, Via R. Cozzi 53, 20125 Milano, Italy.
E-mail: dario.narducci@unimib.it

Claus R. Poulsen
Department of Micro-and Nanotechnology, Technical University of Denmark, Produktionstorvet 423, DK-2800 Lyngby Denmark.
E-mail: claus.poulsen@nanotech.dtu.dk

Noemi Rozlosnik
Department of Micro-and Nanotechnology, Technical University of Denmark, Produktionstorvet 423, DK-2800 Lyngby Denmark.
E-mail: noemi.rozlosnik@nanotech.dtu.dk

Josep Samitier
Institut for Bioengineering of Catalonia (IBEC). C/Baldiri Reixac 10-12, 08028 Barcelona, Spain.
E-mail: jsamitier@ibecbarcelona.eu

Marta Sanmarti
Institut for Bioengineering of Catalonia (IBEC). C/Baldiri Reixac 10-12, 08028 Barcelona, Spain.
E-mail: msanmarti@ibecbarcelona.eu

Ji Wook Shim
Department of Electrical Engineering, Stinson-Remick Hall, University of Notre Dame, Notre Dame, IN 46556.
E-mail: Jiwook.Shim.5@nd.edu

Viktoria V. Shumyantseva
Institute of Biomedical Chemistry, Russian Academy of Medical Sciences, Pogodinskaya St. 10, Moscow 119121, Russia.
E-mail: viktoria.shumyantseva@ibmc.msk.ru

Joon Myong Song
Research Institute of Pharmaceutical Sciences and College of Pharmacy, Seoul National University, Seoul 151-742, South Korea.
E-mail: jmsong@snu.ac.kr

E.V. Suprun
Institute of Biomedical Chemistry, Russian Academy of Medical Sciences, Pogodinskaya St. 10, Moscow 119121, Russia.
E-mail: lenasuprun@mail.ru

Winnie E. Svendsen
Department of Micro and Nanotechnology, Technical University of Denmark, Building 345 east, Lyngby 2800, Denmark.
E-mail: winnie.svendsen@nanotech.dtu.dk

Shutao Wang
Beijing National Laboratory for Molecular Sciences (BNLMS), Institute of Chemistry, Chinese Academy of Sciences, Beijing, 100190 (P.R. China).
E-mail: stwang@iccas.ac.cn.

1

Biased Diffusion and Rectified Brownian Motion at the Nanoscale Driving Mobile Sensing Automata for the Early Diagnosis of Endogenous Diseases

Dario Narducci[1,a,*] and *Gianfranco Cerofolini*[1,b]

INTRODUCTION

Early detection of pathological conditions is known to make therapy much more effective in a number of situations. As an example, pancreatic cancer is the fourth leading cause of cancer-related death in both men and women, although accounting for only about 2.5% of the total neoplastic diseases. This is well known to be related to the absence of non-ambiguous markers enabling early detection of the illness prior to its systemic stage. Median survival from pancreatic cancer diagnosis is around 3 to 6 mon, with a 5-yr survival rate of less than 5%.

Techniques to improve detectability of rare circulating tumor cells have been proposed in the literature over the last few years, some of which make use of the new possibilities opened by micro- and nanotechnology. Among them, the use of microchip technologies somewhat correlates to our

[1]Department of Materials Science, University of Milano Bicocca, via R. Cozzi 53, 20125 Milano (Italy).
[a]E-mail: dario.narducci@unimib.it
[b]E-mail: gianfranco.cerofolini@mater.unimib.it
*Corresponding author

approach. Based upon the evidence that viable tumor-derived epithelial cells (i.e., circulating tumor cells) could be identified in peripheral blood from cancer patients, it was suggested their detection could act as an alternative to biopsies for non-hematologic cancers (Nagrath et al. 2007). Using a microfluidic lab-on-chip capable of separating circulating tumor cells from peripheral whole blood samples, they could identify tumor cells in patients with metastatic lung, prostate, pancreatic, breast and colon cancer with a detection limit down to 5 cells/cm^3. They could also show detection of rare circulating tumor cells in patients with early-stage prostate cancer.

Such an early detection strategy could be extended to enable mass screening, if biomarkers (*sensu lato*) were detectable without the need of sampling tissues. Furthermore, a number of detectors were either distributed or moving through the circulatory system, detection *frequency* (i.e., the probability of detecting a target marker) would significantly increase. In a previous publication (Cerofolini et al. 2010) showed that inoculated swarms of automata with a sectional area of order 10–100 μm^2 (i.e., comparable to that of erythrocytes) can be devised, acting as an early warning system toward endogenous diseases. Such a view which is neither unphysical or simply speculative may become clearer considering that nanoprobes with a size in the 10-nm range can be currently built enabling detection of metabolites (Narducci 2011). At the same time, technology exists to manufacture information processing devices with a component density in the range 10^2–10^3 μm^{-2}, making 10–μm^2 automata powerful enough to collect and process the information provided by their nanoprobes.

In this chapter the key issues connected with making of nanometric sensing automata will be discussed, showing how they can be linked to the immune system in the detection of endogenous pathological conditions. Special emphasis will be given to the motility issue. Since this chapter is mainly addressed to physicians, biologists and health-care specialists, we will dedicate the preliminary discussion of the (rectified) Brownian motion in contemporary physics. Based on these premises, motor design built upon biased diffusion and the so-called rectified Brownian motion will be presented, discussing it in view of the unusual aspects of propelling objects in fluid at the nanoscale. The additional advantage stemming from rectified Brownian motion of being *self-directive* will be explained. Actually, the dissipation-fluctuation phenomenon may orient nanobots with respect to chemical gradients. Use of such a property to address automata toward malignant cells and tissues will be discussed and compared to standard nanodiagnostics.

FROM NANOPARTICLES TO NANOBOTS

The blood circulatory system is able to bring in the proximity of all cells (i.e., at a distance short enough to allow O_2 and glucose to diffuse from blood to the cells) any particulate with a size lower than the red blood cells. The O_2 diffusion distance is around 50 μm (Brown 2000) while the red blood cell has a mean diameter close to 5 μm.

Particulates can leave the circulatory system entering the tissue parenchyma only if they are smaller than the capillary pores (with diameter of less than 1 μm), but they can wander through interstitial (i.e., intercellular) water only if their size is below 0.1 μm. The exploration of the parenchyma therefore requires very small particles; a lower limit to particle diameter is however determined by renal clearance, which is highly efficient to reduce the blood lifetime on the hour time scale when the particle diameter is below 0.01 μm (Choi et al. 2007). Thus only particles with diameter in the interval 10–100 nm are able to reach as a whole in each part of the body through the circulatory and lymphatic systems and through the interstitial water. The medical interest toward such *nanoparticles* (NPs) is due to the fact that they can be conjugated to antibodies or peptides at their surface to bond cancerous cells, thus providing a tool for an early-stage imaging of the tumor (O'Farrell et al. 2006). An additional conjugation to antineoplastic agents (cancer chemotherapic drugs, cytotoxic drugs) thus seems a way for their selective and efficient delivery. With this aim, several materials have been considered, ranging from polymers (Soppimath et al. 2001) and liposomes (Weinstein et al. 2010) to carbon clusters (Berlin et al. 2010). As an alternative, NPs may operate as physical devices conveying heat on tumoral cells (Pankhurst et al. 2003).

In vitro and *in vivo* studies based on this hypothesis have given very promising results. With the birth of nanomedicine (i.e., the use of nanoscale or nanostructured materials) that according to their structure, have unique medical effects (Wagner et al. 2006) and may be considered to go back just to the application of NPs for diagnosis and therapy.

Clearly enough, use of NPs is the simplest possibility of exploiting the power of nanotechnology in medicine. NPs are simple to make, widely available for a number of chemicals and relatively cheap to manufacture. However, from differing points of view they are far from being the ideal tool even for imaging and therapeutic applications. In both respects NPs are actually carriers of chemicals. Thus, their effectiveness relates to the large surface-to-volume ratio they exploit compared to macroparticles. However, geometry tells us that spheres are solids with the lowest surface-to-volume ratio for any given size D. Stated differently, while for any solid the surface-to-volume ratio scales as $1/D$, so that the transition from macro to nano enables large surface availability for a given total volume, for any given *D* the sphere is the solid characterized by the minimum surface. Geometry does not indicate a solid with the maximum surface for a given *D*—but anything not spherical would be an advantage. In this sense, the increasing number of techniques available these days to obtain nanometric quasi-one-dimensional structures are expected to shift the current paradigm from nanoparticles to nanowires.

Nanowires (NWs) are almost cylindrical objects with a diameter in the interval 10–10^2 nm and much larger length, typically > 10^3 nm (Figure 1.1). The medical use of NWs is essentially due to their extraordinary detection potentials. In recent years a huge interest has been shown on functionalized silicon nanowires to be used as sensing probes in different classes of biosensors (Loncar et al. 2006). Nanowires actually enable a gigantic extension of the active surface, with a consequent improvement of the sensitivity and of the detection threshold down to the part-per-billion range. Use of nanowire-based sensors was initially proposed by Lieber's group (Cui et al. 2001) for inorganic sensing, and further exploited since then by several research groups (see, e.g., (Zhou et al. 2003, Patolsky et al. 2006a), and citations therein). In most cases, nanowires are used to build NanoFETs, namely as bridges between two electrodes acting as the Field-Effect Transistor (FET) channel (Patolsky et al. 2006b)—although different layouts were also proposed (Stern et al. 2007). Upon adsorption, carrier depletion—accumulation modulates the nanowire resistance. Lieber showed how biotin-modified nanowires could detect streptavidin down to the picomolar concentration range (Cui et al. 2001). With the same design, antigen-functionalized silicon surfaces were proved to reversibly detect antibodies with response times of the order of the second. Silicon nanowires were also used to obtain highly sensitive and sequence-specific

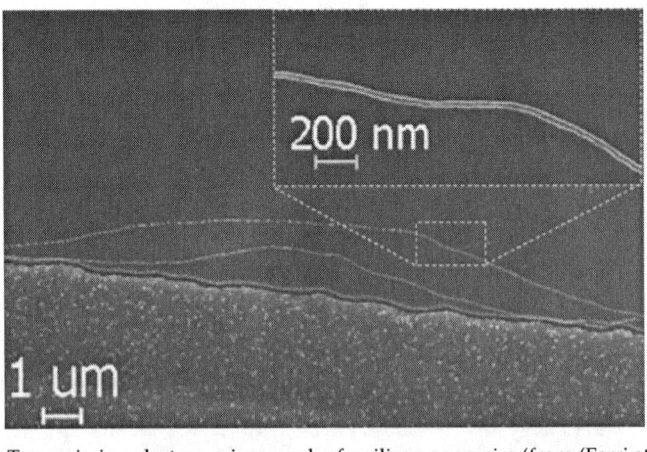

Figure 1.1: Transmission electron micrograph of a silicon nanowire (from (Ferri et al. 2011)).

conductimetric sensors by immobilizing single stranded probe DNA molecules on the nanowire surfaces. Sensitivities down to 25 pM of target DNA in solution could be achieved (Li et al. 2004). Two-terminal silicon nanowire sensors could further be used as selective detectors of DNA. The surface of nanowires was modified with peptide nucleic acid (PNA) receptors designed to recognize wild type vs. the ΔF508 mutation site in the cystic fibrosis transmembrane receptor gene (Hahm et al. 2004). Also in this case, the silicon surface was previously terminated with avidin, that acted as the linker to allow the immobilization of PNA (the actual sensing probe). Nanowires were also integrated into arrays for medical sensing. Multiplexed detection of four prostate cancer markers (prostate specific antigen (PSA), PSA-α1-antichymotrypsine, carcinoembryonic antigen, and mucin-1) was achieved by using a set of silicon nanowires. The silicon surface was modified by assembling aldehydes that were then used as linkers to immobilize the relevant antibodies (the bioreceptors), suitably terminating the unreacted aldehydic groups (Zheng et al. 2005). The detection limit was found to be in the range of a few picograms per cubic centimeter.

Nanowires (as well as nanoparticles) are good examples of how nanotechnology can positively impact on medicine. However, by and large they appear to be just appetizers. In many senses these nano-objects are exploitations of the least revolutionary nanotechnological feature, namely that of large areas—and some enhanced capability of *in vivo* diffusion through biological barriers. Nanotechnology, however, has a host of additional features to deliver, one is the possibility of designing complex devices on a microscale. The transition of the microelectronic rule of design to the nanometric scale actually enables the possibility of devising (realistically!) automata of micrometric sizes able to perform *in vivo* complex tasks, including the detection of specific metabolites and the treatment of pathologies at the cellular level. With the advantage of an intrinsic capability of navigating the body.

SWARMS OF NANOBOTS: OUTLINE OF AN ACTUAL STRATEGY OF BIODETECTION

In this chapter we will use the word *nanobot* to indicate any artificial machine with overall size of the order of a few micrometers or less in all spatial directions, constituted by nanoscopic components with individual dimensions in the interval 1–10^2 nm, and able to perform sophisticated functions such as navigation, recognition, data collection and transmission (Requicha 2003). In view of their limited size, nanobots are essentially devices with very limited computational resources and their interaction can only be local. These facts suggest that the algorithms for their control should be based on swarm intelligence (Bonabeau et al. 1999). The idea of using swarm of nanobots is not new. Methods that use self-assembling agents are discussed in Requicha (2003) and Arbuckle et al. (2004) while medical applications of nanobots are described in Cavalcanti et al. (2005), Freitas (2005), Cavalcanti et al. (2006), Freitas (2006), and Hogg (2008). In particular, chemical communication techniques for nanobots foraging in fluid environments relevant for medical applications are discussed in Cavalcanti et al. (2006), while the problem of designing control systems for nanobots is analyzed in Cavalcanti et al. (2005).

At the present stage of knowledge, the hypothesized swarm is certainly far from being producible, but it is not a chimera because most of the critical steps required for its preparation have already been established.

A nanobot can be endowed with a limited amount of intelligence, however allowing it to master its numerous and sophisticated functions (sensing, recognizing of the chemical pattern, controlling the motion, managing the power, and so on) (Cerofolini et al. 2010). A 100-kbit circuitry is expectedly able to manage the information coming from a few (of the order of 10) sensing regions, each specialized to the identification of a different metabolite. Starting from the reasonable assumption that in the next 20 years integrated circuits will attain a density on the scale of 10^{11} cm^{-2} and that the nanobot will be fabricated employing the planar technology, hosting 10^5 devices requires an area of just 10^2 μm^2.

Assuming that specialized portions of the electronic part of the nanobot (the gates of selected transistors) are in contact with the external world and so functionalized that the adsorption of a specific molecule affects the transconductance of the underlying FET by an amount so large to allow the detection of the phenomenon. In this case the considered transistor could be used as a NanoFET sensor.

The integration of computation and chemical sensing in the same circuit, however, is not easy. In fact, logic and memories are built in such a way as to decouple as far as possible the electronic device from the outer world. The basic constituents of the electronic circuit, the FETs, are usually buried beneath approximately 10–20 dielectric and metallic layers of total thickness of a few micrometers; if the transistor is produced via the silicon-gate technology, the exposure to the outer world of the FET core—the gate oxide—is even conceptually impossible. On another hand, a new paradigm based on the crossbar structure has been proposed (Cerofolini et al. 2007, 2008, 2009). The crossbar allows a cross-point density even larger than the one achieved by standard FETs, may be embedded in microelectronic circuitry, can be built in the back-end stage of the fabrication process, and can thus be easily put in contact with the environment.

Biomimetic coatings designed not to activate the immune response of the organism are also within reach of current technology, as demonstrated in the case of lipid monolayers or bilayers (mimicking biological membranes) supported on solid or polymer surfaces on even larger areas (Morigaki et al. 2002, An et al. 2006, Cashion and Long 2009). Of course, the supported films are only weakly bound and are therefore not very stable; however a larger stability could be achieved by bonding covalently (e.g., via silanization) to the sensor surface molecules with carboxylic terminations, mimicking the outer surface of cells. Alternatively, another route involving the hybridization of the nanobot with the membrane of red cells seems within reach of near-future technology.

However an additional issue needs to be considered, namely that of supplying power to the nanobot. Power is needed at least for three functions: sensing, processing and transmitting information, and propelling. While this may appear unusual, powering is the most critical issue when it comes to devising nanobots. Actually, none of the standard solutions one might think to may be downscaled to the micrometric (or submicrometric) scale. Batteries are very complex devices to retain their efficiency at the nanoscale, and the energy they are capable of storing is anyway too small to be of use. Local power generation is an interesting alternative that may be considered. Power has already been successfully generated at the submillimetric scale using thermoelectric or microelectromechanical (Williams et al. 1996, Jeon et al. 2005) devices. Micro- or nano-dynamos (reverted electrical micro- or nanomotors) might also be considered. In this case however, it is probably not too far from reality to suggest that a sparing energy policy is a must. While a nanobot ubiquitously propelled through the body and continually sensing and elaborating information would set a quite unsolvable problem of power consumption, a wiser use of energy might alleviate the issue of local powering. In principle, the nanobot computational capabilities might be made idle while the automaton is neither sensing or communicating. However, if the sensing ability was considered as an independent property, it would be difficult to imagine a protocol cleverly turning the sensor on and off. As a result, some of the computational engine should also be kept working at all times—unless it is stipulated to reduce the nanobot vigilance by lowering its sampling frequency while moving through the body. While at first sight propulsion might appear not to be essential as nanobots may simply move with the blood stream, sensing efficacy might be greatly improved if one could implement an additional capability for the nanobot to locally address its navigation in response to a variation of a biomarker concentration. We will show that sensing and propelling can be instead coupled at a fundamental level—namely setting a propulsion mode such that the motion occurs as a result of chemical gradients, leading the device to head at chemical sinks (locally) not as a result of a retroaction ('sense, then follow') but as an effect of biased diffusion in the presence of a chemical gradient. This integration of sensing and propulsive capabilities cuts to zero the energetic cost of two expensive functionalities—and re-enables the possibility of keeping most of the computing electronics idled for most of the time.

BIOSURVEILLANCE AND COOPERATIVE BEHAVIOR

Thus, nanobots may be thought of as nanolaboratories cruising through living organisms and enabling detection of pathologies at their earliest stage of development, actually dispatching notices about the health status of the organism to some external receiver. The onset of these capabilities as *biosurveillance* will be referred to here. While it has be shown (Cerofolini et al. 2010) that current technology is insufficient to enable biosurveillance at the gene or cell level, it can be proved instead that this is feasible at the circulatory-system level. Indeed, a nanobot in a capillary could feel the metabolic pattern of the family of cells fed by the capillary itself, thus surveying the cells contained within a diffusion length of the biomarker. For the biosurveillance to occur, residency time of the nanobot in the vicinity of the cell must be not shorter than the time needed for sensing. A dimensional argument allows in estimating that each nanobot drifting with the blood flow will visit a cell every 0.3–0.6 s—enough for a nanosensor to detect and quantify the target biomarker. Furthermore, since blood makes a whole cycle through the body about every 60 s, each nanobot will explore approximately 10^5 capillary per month. Thus, a swarm of just 300,000 nanobots will ensure each cell to be sensed 10 times a month.

An additional requirement for effective sensing is that the sampled blood volume eventually be large enough to contain at least one biomarker molecule. While this is likely to happen when macroscopic samples of volume ω (e.g., 1 cm^3) out of a total blood volume Ω of 5×10^3 cm^3 are considered, the issue is much more critical when sampling involves nanoliter volumes. Actually, the smallest number N_{thr} of biomarkers in the human body statistically ensuring one biomarker molecule in the volume ω can be computed using a Poisson distribution. Thus, $N_{thr}\omega/\Omega \approx 1$, so that for $\omega^{1/3}$ of order the biomarker diffusion length, i.e., $\omega \simeq (10^2\ \mu m)^3 = 10^{-6}\ cm^3$, one gets a detection threshold N_{th} of about 5×10^9. Such a number can be however lowered if the nanobots may be thought as *cooperating* devices, possibly with the additional capability of directed motion. Cooperation from this standpoint simply implies that their capability of detecting biomarkers is not simply the result of their individual ability to find and detect them in a certain basin of the body—but is amplified by the additional functionality of signaling to other swarm members in the regions where markers have been found. To this aim, a *directive* approach (namely one where actual messages are exchanged and processed by individual units) is easier to think but almost impossible to implement, since this would require an expense of energy and an on-board intelligence that are beyond the reach of current technology. Instead, spontaneous swarm organization might offer an interesting alternative. Let us suppose that a capillary of section d feeds a tumoral cell (or anyway a cell releasing a targeted biomarker). Let us also assume that nanobots of size d_n patrol the system, and that they are suitably engineered to dock onto malignant cells. Since $d < d_n < 2d$, the first nanobot sensing the marker will adhere to the cell, locally reducing the capillary section (although not obstructing it). Therefore, the blood flowing rate will locally increase, also increasing the number of nanobots visiting the capillary per time unit, although incoming nanobots will be unable to overcome the first, docked unit. As a result, the surroundings of the malignant cell will be more thoroughly explored and controlled, and this in turn will increase the chance that other nearby malignant cells be detected—*as if* the first unit had signaled to its swarm a critical situation.

In summary, it can be seen in this section that nanobots are the natural, physically and technologically feasible evolution of the present NP-oriented approach of nanomedicine. We have shown the rather obvious advantages of deploying smart, cooperating nanobots to patrol a living organism instead of simply deploying passive, dumb particles—at the most able to deliver biomolecules with some specificity or to bind to specific tissues. We have discussed many issues on implementing nanbots for biosurveillance. However one point remains which is strictly related to manufacturing. In order for a technology to be such, a sufficient amount of devices must be available to serve at least a niche of users. Translated into a medical language this implies answering two questions, namely (1) how many nanobots are needed to effectively patrol a patient; and (2) do we have a manufacturing technology consistent with this need? The first question may be answered focusing on a specific application (of the many addressable)—early recognition of tumors. The circulatory system of an adult human contains approximately 2×10^{10} capillaries, feeding ≈ 3,000 healthy cells each. Taking into account that the circulatory system is optimized and thus unable to feed a much larger number of cells, it is not surprising (and is well known) that the metabolism of cancer cells, ≈ 50 times faster than that of healthy cells, is self-limiting. A capillary being able to feed approximately 3,000/50 = 60 cancerous cells. Thus the growth of the *tumorlet* is hampered by the insufficient flow of oxygenated blood to the cancerous cells, ≈ 98% of which necrotize by heavy tissue hypoxia. This is true up to the *angiogenic switch* (Sherwood et al. 1971), when the tumor itself promotes a neo-vascularization able to sustain its subsequent explosive growth and its transformation from a local disorder into a systemic diseases (Welter and Rieger 2010). Until the angiogenic shift has occurred (that does not necessarily happen) the tumor is referred to as in a *dormant phase*. The tumor remains in the dormant phase up to a size of 1 mm size for which the cancerous non-necrotized cells therein are about 2×10^4 (2% of the number of cells in 1 mm^3). After the angiogenic switch and when tumor imaging is currently possible (size $\geq$ 5 mm), the tumor contains up to 10^8 cells. An effective biosurveillance should be able to detect a tumor at its earlier stage. This implies that the number of nanobots patrolling the circulatory system (and the whole body thereof) should enable all capillaries to be visited within the clearance time of the nanobot swarm. Let us assume a nanobot clearance time of 24 hr. Since each nanobot (as each blood volume) performs an entire cycle through the circulatory system in about 1 min, the number of cycles allowed for the automaton is 1,440. It can be further estimated that the number of capillaries in an adult is of $\approx3.5\times10^5$, the average length of each being around 250 μm (Cerofolini et al. 2010). This totals to a length of about 10^9 cm. At an average blood speed of 5×10^{-2} cm/s, the fraction of capillaries visited by each nanobot before being expelled from the organism is

$$\frac{\left(5\times10^{-2}\,\mathrm{cm/s}\right)\times\left(60\ \mathrm{s}\right)\times1440}{10^{9}\,\mathrm{cm}}\approx4\times10^{-6}$$

The minimum number of nanobots required to give a probability close to 1 to visit each capillary is thus 2×10^5.

Although this figure may appear huge and discouraging,to some extent it is not. Currently the microelectronic technology is able to make some 10^9 transistor out of a silicon wafer. Thus, a technology compatible with microelectronic standards should be able to provide about 5,000 'doses' of nanobots per processed wafer—confirming also in this respect the overall feasibility of this type of *active nanotechnology* applied to medicine. It has however to be mentioned that technological feasibility is a necessary but not sufficient condition for a technological idea to become a technology. Especially in the case of a forefront application, it is usually preferable to introduce it to the end-user as an evolution of a traditional approach, privileging the implementations with lower complexity and costs. In this case, an exploitation bridging NPs and nanobots might be desirable. As an example, nanowires able to navigate through the body, binding selectively to cancerous cells and delivering some therapy might be a reasonable example. From the viewpoint of the nanotechnologist, this would be an excellent chance to test the possibility of directing NWs toward cancerous cells using the tool of biased diffusion, that can be implemented in low-symmetry objects only. In the next section we will discuss this exciting possibility in greater detail. To conclude it may be worth mentioning that the medical pros of using NWs instead of NPs would anyway be in the largest number of biomolecules deliverable per mass unit of nano-objects (*geometric advantage*) and in the possibility, descending from biased diffusion, of operating with lower doses of nanodrugs (*invasiveness advantage*).

THE PHYSICS OF STANDARD AND RECTIFIED BROWNIAN MOTION

Brownian motion gets its name from the observation made by Robert Brown of the erratic motion of pollen particles in a water suspension. The phenomenon can be simply explained by using some mathematics as the effect of the collisions of water molecules on the pollen particles. Let M and T be the particle mass and the absolute temperature, resp., and let us assume that particles are also subject to viscous friction in the medium (water as a fluid), damping their motion as the effect of a force $\mathbf{F}_v = -\lambda \mathbf{v}$ (where λ is the viscosity coefficient and $\mathbf{v}$ is the particle velocity). Thus, under equilibrium conditions, the instantaneous force $\mathbf{f}(t)$ acting on each particle must be subject to the following constraints: (1) its time-average has to be isotropous, i.e., $\int_{-\infty}^{+\infty} \mathbf{f}(t)\,\mathrm{d}t = 0$, otherwise a net force would be acting on the system leading to a net particle flow, which opposes the equilibrium hypothesis; (2) it must fluctuate in time, i.e., $\int_{-\infty}^{+\infty} \mathbf{f}(t) \cdot \mathbf{f}(t')\,\mathrm{d}t = \Lambda \delta(t - t')$ (where Λ is a constant and $\delta(x)$ is the Dirac function). Thus, according to the second law of mechanics,

$$M\mathbf{a}(t) = M\frac{\mathrm{d}\mathbf{v}(t)}{\mathrm{d}t} = -\lambda \mathbf{v}(t) + \mathbf{f}(t) \tag{1}$$

(where $\mathbf{a}$ is the acceleration) that integrates to

$$\mathbf{v}(t) = \mathrm{e}^{-\lambda t/M}\left(\int_0^t \mathrm{e}^{-\lambda t'/M}\frac{\mathbf{f}(t')}{M}\mathrm{d}t' + \mathbf{v}(0)\right) \tag{2}$$

so that it is easy to verify that

$$\int_{-\infty}^{+\infty} \mathbf{v}(t)\,\mathrm{d}t = \mathbf{v}(0)\mathrm{e}^{-\lambda t/M} \tag{3}$$

this implies that the average velocity falls to zero within a time of order $\tau = M/\lambda$. Typically, $M \approx 10^{-13}$ kg while $\lambda \approx 10^{-7}$ kg/s so that $\tau \approx 10^{-6}$ s, meaning that no directional motion survives more than a few microseconds. The displacement $\mathbf{r}(t)$ is obtained by integration of Eq. (2):

$$\mathbf{r}(t) = \int_0^t e^{-\lambda t'/M} \left(\int_0^{t'} e^{-\lambda t'/M} \frac{\mathbf{f}(t'')}{M} dt'' + \mathbf{v}(0) \right) dt' \tag{4}$$

leading to a time-averaged value of zero for any time t significantly larger than τ, as expected. This is consistent with the second law of thermodynamics, since no work can be extracted from a system in the absence of temperature differences.

The situation drastically changes when the system is kept out of equilibrium (Astumian 1997), which is the typical situation one encounters in biological systems. The first example of Brownian motors in the literature goes back to the famous paradox of Feynman's ratchet (Feynman et al. 1966). In its most interesting representation, Feynman's ratchet is a microscopic wheel with asymmetric teeth bound to a paddled shaft (Figure 1.2). The paddles are in contact with a gas at temperature T_1 while the ratchet itself is at a temperature T_2. Due to the mechanism limiting the free rotation of the wheel, one may expect that collisions of gas atoms (or molecules) with the paddles should impart the wheel a rotation—setting up something that, at first sight, might look as a perpetual motion machine of the second kind. While this would be actually the case when $T_1 = T_2$, when $T_1 \neq T_2$ the motor is found to work as intuitively expected, the wheel rotating either clockwise or counterclockwise. Feynamn's ratchet is actually paradoxical only when the whole system is at the same temperature. In that case, would the shaft rotate, the device would extract mechanical work out of a heat sink at a uniform temperature, in contrast with the prescription of Carnot's theorem that predicts a null conversion rate for any thermal

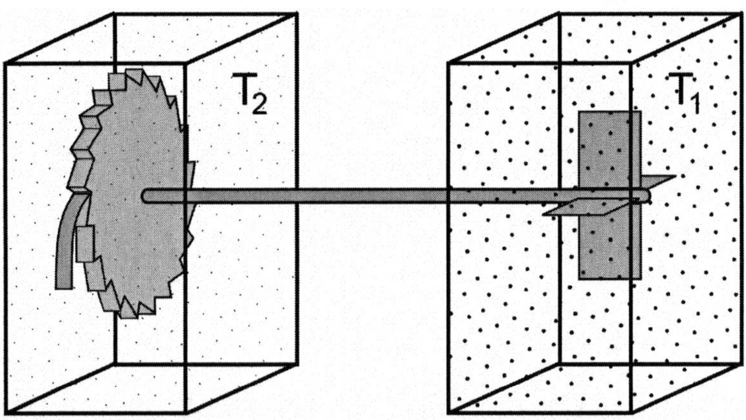

Figure 1.2: Sketch of a Feynman's ratchet. Note that paddles and ratchet have to be at different temperatures for the paddles to rotate.

engine operating between two heat sinks at the same temperature. Instead, when the engine operates between two sinks at different temperatures, conversion of heat into work violates no thermodynamic law, since energy is conserved (heat is converted into mechanical work—first law) and part of the heat is not converted into mechanical work but is simply transferred as such from the hot to the cold sink (second law). It can be actually seen (Astumian et al. 1998) that the net rotation rate ω depends upon T_1 and T_2 as:

$$\omega \propto \frac{\Delta U(T_2 - T_1)}{2k_B((T_1 + T_2)/2)^2} \exp(-2\Delta U / k_B(T_1 + T_2)) \tag{5}$$

where ΔU is the potential barrier the tooth has to overcome for the wheel to rotate. Note that $\omega = 0$ for $T_1 = T_2$, as expected, while ω gets to a maximum when the average temperature $(T_1 + T_2)/2$ equals $\Delta U/k_B$. Also note that ω ¤ 0 as T_2 ¤ T_1—namely the ratchet may either rotate clockwise or counterclockwise depending on the temperature difference between the paddles and the cog and irrespective of the tooth shape. Although devised as an educational *gedankenexperiment*, Feynman's ratchet is a proper embodiment of a motor based upon rectified Brownian motion. It actually converts the random impacts of atoms (molecules) onto the paddles into the directional motion of a macroscopic device.

Since Feynman's ratchet, quite a bit of theoretical endeavor has been put forward to verify not just the possibility—but also the actual occurrence—of rectified Brownian motion in out-of-equilibrium systems. In 1997 Astumian (Eichhorn et al. 2005) proposed an alternate scenario for a motor able to extract mechanical work out of thermal fluctuations. Suppose we have a set of N particles subject to a potential landscape $u(x)$. For the sake of simplicity we will focus on a one-dimensional system. At $t = 0$ let $u(x) \equiv 0$, i.e., no potential is applied. Thus, if the particle sits *on the average* at $x = 0$, the probability $P(x,t)$ of finding it at x at time $t > 0$ will be $P(x,t) \propto \exp\left(-x^2 / \sqrt{4Dt}\right)$.

Now, if we suddenly turn on a sawtooth potential

$$u(x) = \begin{cases} \dfrac{u_0}{\alpha L} x & \text{for} \qquad 0 < x - jL \le \alpha L \\ -\dfrac{u_0}{(1-\alpha)L} x & \text{for } (\alpha - 1)L < x - jL \le 1 \end{cases} \tag{6}$$

where L is the system period in space, j is a periodicity index ($j = 0, \pm 1, \pm 2, \ldots$), and u_0 and α are positive constants, what will happen is that the number of particles trapped on the left and on the right of the potential crests (at $x = x_{-1}$ and $x = x_1$) surrounding the potential well at $x = 0$ will be different if $\alpha \neq 1/2$. More precisely, the N particles will partition into three subsets (Figure 1.3 (b)):

$$\begin{aligned} N_{-L} &= N \int_{x_{-2}}^{x_{-1}} \exp\left(-\frac{x^2}{\sqrt{4Dt}}\right) \mathrm{d}x \\ N_0 &= N \int_{x_{-1}}^{x_1} \exp\left(-\frac{x^2}{\sqrt{4Dt}}\right) \mathrm{d}x \\ N_L &= N \int_{x_1}^{x_2} \exp\left(-\frac{x^2}{\sqrt{4Dt}}\right) \mathrm{d}x \end{aligned} \tag{7}$$

where the subscripts to the N's label the position of the relevant potential well minima while x_i ($i = \pm 1, \pm 2, \ldots$) enumerate the extrema of $u(x)$ around $x = 0$. Clearly enough, particles will fall down into the potential minima. It is easy to verify that $N_0 > N_{-L} > N_L$ for $\alpha > ½$ (Figure 1.3(c)). Thus, if we turn the potential off again, the distribution will feature three Gaussian curves centered around -L, 0, and L, one of the lateral Gaussian being

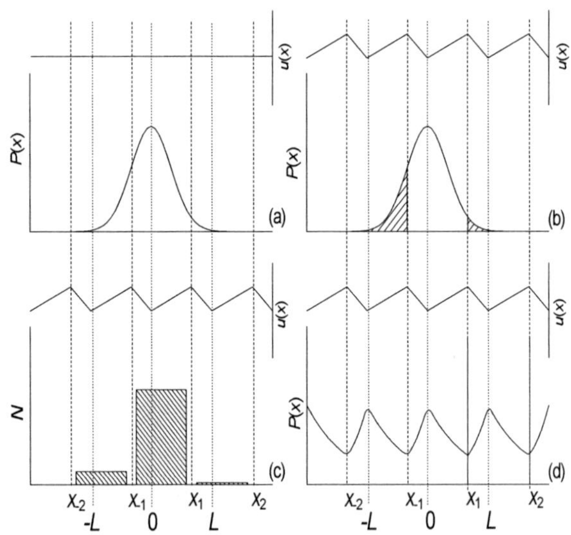

Figure 1.3: Rectified Brownian motion in the presence of a flashing asymmetric potential. (a) At $t = 0$ the particles sit around $x = 0$ while $u(x) = 0$. (b) At $t > 0$ a sawtooth potential is turned on, so that particles are trapped on the left and on the right of the potential crests surrounding the potential well at $x = 0$. (c) Number of particles trapped into each potential well. Note that the distribution is asymmetrical, more particles having moved to the left. (d) Same situation as in (b) but with the potential being turned on slowly enough to allow the system to equilibrate. In this case the population would simply oscillate between a uniform distribution and the Boltzmann prediction, no drift occurring.

larger. Overall, should we inject particles in $x = 0$ and periodically turn on and off $u(x)$ we would generate a net particle flow rightwards ($\alpha > ½$) or leftwards ($\alpha < ½$). This result is noteworthy since no net force is applied to the particles, as $\int_{-(1-\alpha)L}^{\alpha L} u(x)\,dx = 0$. The particle flow is actually the mere result of a diffusive (Brownian) motion in the presence of an *asymmetric potential landscape* bringing the system *out of equilibrium*. Both aspects have to be stressed. The potential needs to be asymmetrical for a directional flow to be observed. Were $\alpha = ½$, it is easy to verify that we would get $N_{-L} = N_L$. Also, it is essential that the flashing potential be turned on and off rapidly enough not to allow the particle distribution to equilibrate (thermalize). Otherwise the particle population would simply oscillate between $P(x) =$ const. (when the potential is off) and $P(x) \propto \exp(-u(x)/k_B T)$ (when the potential is on—Figure 1.3 (d)), and no net flow would be observed.

It is noteworthy to observe that the situation depicted in this simple model, although apparently artificial, is not unrealistic. A flashing asymmetrical potential, further to be externally imposed, may spontaneously generate in a Brusselator (Prigogine and Lefever 2007), where cyclic reactions (e.g., a Belusov-Zabotinski reaction) generate patterns in the concentration maps of chemicals. Thus, if the particles actually sense chemical gradients (e.g., by setting short-range attractive interactions with one of the chemical), they will directionally diffuse. As a second example, let us consider a system kept out of equilibrium by the cyclic variation of a chemical concentration. Oxygen in the blood stream is a possible example. We also stipulate that, at any given time, such a concentration is not uniform in space. To this aim, we may assume for instance that a solution containing a chemical S flow in a two-dimensional channel (Figure 1.4), and that the rate of consumption of S at the top and bottom channel walls be different: $r_{top} \neq r_{bottom}$. Note that the solvent has virtually no momentum normal to the channel axis. Thus, a particle put in the fluid will be swept away by the solvent stream, moving along the channel axis. However, should the particle be capable of some privileged interaction with S, things will appear different. Suppose for instance that the collisions between the particles and the molecules of S be completely inelastic. Thus, while the collisions with the solvent molecules will simply generate a standard, non rectified Brownian agitation normal to the axis channel (cf. Eq.(4)), collisions with the S molecules will cause a net drift in the normal-to-the-axis direction. After each collision the final velocity $\mathbf{v}_f$ of the particle will be

$$\mathbf{v}_f = \frac{m_p \mathbf{v}_p + m_S \mathbf{v}_S}{m_p + m_S} \tag{8}$$

(where m_p and m_S are the masses of the particle and of the S molecule, resp., while $\mathbf{v}_p$ and $\mathbf{v}_S$ are their velocities before the collision). Since $m_p \gg m_S$, then $\mathbf{v}_f \approx \mathbf{v}_p + (m_S / m_p)\mathbf{v}_S$. Thus, the normal-to-the-axis component of the particle velocity $v_{p\perp}(t)$ accounts to

$$v_{p\perp}(t) = v_{p\perp}(0) + \frac{m_S / m_p}{\tau_S} \int_0^t v_{S\perp}(t')dt' \tag{9}$$

where τ_S is the average time between two collisions. Since $v_{p\perp}(0) = 0$ but $\int_0^t v_{S\perp}(t')dt' \neq 0$ we will observe a drift of the particle toward the wall with the highest demand for S. Figure 1.4 shows the actual chemical fluxes taking $r_{top}/r_{bottom} = 0.1$.

A third and final possibility should be mentioned, although too complex to be analyzed in detail in this chapter (Eichhorn et al. 2005). Whenever interactions among particles are found, in a out-of-equilibrium asymmetrical system we can observe a wealth of intriguing, counterintuitive phenomena that occur. Further to rectified Brownian motion, particle mobility may exhibit either an absolute negative mobility —namely moving in the opposite direction with respect to a (weak) applied force; or a differential negative mobility, moving according to the applied force but with a mobility decreasing with the applied force, acting as a 'Brownian donkey'.

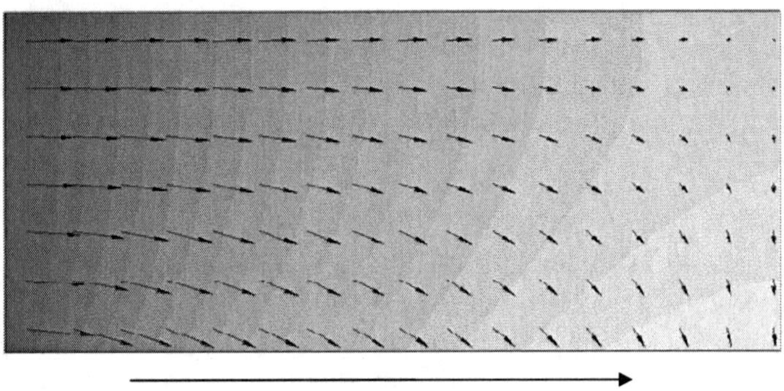

Figure 1.4: Preferential radial flow as induced by an asymmetric chemical demand by tube walls.

The main conclusions of this section are summarized below. We have seen that, out of equilibrium, the presence of an asymmetrical potential (either periodic or not) may lead to the transformation of thermal noise into mechanical work. Although the examples considered here are not more than toy models, these results are strongly suggestive in view of the twofold possibility of explaining motility *in vivo* and to devise machines able to self-propel in biological systems. In living matter, non-equilibrium is actually the name of the game, so the first condition for rectified Brownian motion is always fulfilled. What we have to imagine (and to make up) is the asymmetric potential landscape. Nature knows how to do it, as will be shown shortly—so that the simplest way to build Brownian motors is to copy what is known to work out of millennia of evolution.

BROWNIAN ENGINES: PROPEL AND DRIVE

Possibly the first detailed account of rectified Brownian motion *in vivo* was the 1998 seminal paper of Fox (1998). In his article, Fox analyzed a rather large number of biological systems of increasing complexity, from ubiquinone diffusion up to muscle fiber dynamics. Although the mathematical complexity of his analyses limited the readership of his paper, his insight into biological motility motivated an increasing number of speculative research efforts, mostly computational for almost a decade. The basic idea behind Fox's models can be easily reviewed referring to the case of ubiquinone reaction-diffusion. Neglecting the complexity of mitochondrial processes (the so-called Mitchell cycle), ubiquinone molecules undergo a coupled (chemical *plus* physical) process (Figure 1.5):

(1) they are reduced or oxidized on the two opposite sides of the cell membrane:

$$UB + 2H^{+} + 2e^{-} \text{ € } \quad UBH_2 \tag{10}$$

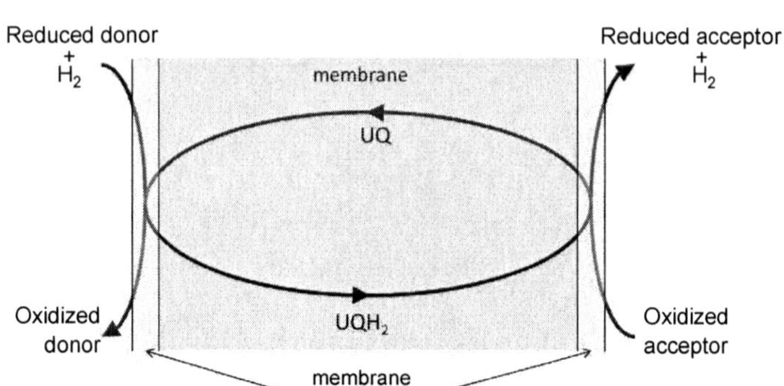

Figure 1.5: Schematics of the ubiquinone reaction–diffusion process at the membrane walls. See text for details.

(2) they diffuse through the membrane in either oxidized (UQ) or reduced (UQH2) form. Correspondingly, protons also diffuse through the membrane *as charged particles.*

In the physiological state, a transmembrane electrochemical potential ΔE_{H^+} is established by energy metabolism, accounting to

$$\Delta E_{H^+} = \Delta\psi - 2.3\frac{RT}{F}\log\frac{[H^+]_2}{[H^+]_1} \quad (11)$$

where $\Delta\Psi$ is the transmembrane electrical potential, R is the universal gas constant, F is the Faraday constant and $[H^+]_i$ are the proton concentrations at the two sides of the membrane. It can be computed that the outward transport of the proton requires some 200 meV. In the standard view, this energy is provided by the redox free energy. Instead Fox showed that the redox free energy is actually spent to create an asymmetric potential landscape, leaving upon rectified Brownian motion (i.e., to thermal fluctuation) the duty of shuttling protons across the membrane. In essence, if the system spends its energy to keep electrochemical potentials consistent with the reduction and the oxidation reactions to occur at the pertinent membrane walls, then stochastic motion brings UBH_2 to the oxidizing wall and UB to the reducing wall, establishing a perpetual flow across the membrane. In turn, this forces protons to climb the potential landscape. The whole picture can be seen with the eyes of the model toys of the previous section. Metabolism keeps the system (the membrane) out of equilibrium and, at the same time, generates an asymmetrical potential (through the ubiquinone densities at the membrane walls). Then Brownian motion (of the proton) therefore gets rectified, enabling its diffusion to be directed across the membrane *against* the electrical potential.

In this light a rather large number of physiological processes have been reinterpreted since Fox's paper . More recently, artificial exploitations of rectified Brownian motion have been reported (Mahmud et al. 2009). In this case, a patterned surface has been made, playing the role of the chemical gradient. The surface is patterned in such a way to act as a ratchet, disabling backward motions of 'rigid' cells. Instead, when the cell can modify its geometry, standard (non-directional) diffusion occurs. Cells of different types were left to crawl onto the surface, showing that the mechanism enables its separation—namely at the same time it guarantees motility and biorecognition.

It may be worth noting that in the case of cells crawling onto the surface, asymmetry is seen by one type of cell but not by the other. The 'rigid' cell recognizes the surface as being covered by asymmetric features while the 'flexible' one does not (almost) discern the textured surface from a flat one. Thus, the latter cell diffuses isotropically while only 'rigid' cells give rise to biased diffusion. This experimental evidence is quite suggestive for nanobot motility. In a biological medium we can easily imagine a large number of different chemicals setting up gradients in any possible direction. However, if the nanobots just *sense* one of them, they will eventually move following (or opposing) that gradient. To make the picture more realistic, consider once again the second toy model we put forward. Suppose that the nanobot outer surface is covered by molecules binding to, say, oxygen (O_2). This mimics quite acceptably the fully inelastic collision of the model. Thus, in the presence of an oxygen gradient we might expect a preferential flow toward the wall with the highest demand for oxygen.

Although interesting, this preferential diffusion is not however sufficient to propel the nano-object. At this stage blood flow still largely dominates its dynamics. If an effective Brownian motor has to be devised, it should be able to actually drive its cargo, not just to deviate its trajectory. This can actually be realized by putting together the ingredients needed to this purpose—namely (1) the motor must 'bring on board' the asymmetric potential landscape needed to exploit rectified Brownian motion; (2) the potential must be asymmetric as of a given chemical only, leading to a motion directed with respect to, for instance, a biomarker or any molecules overexpressed by malignant cells; and (3) the ratchet it builds up must have a reasonable efficiency, as it has to be effective even in the presence of a conventional source of mechanical momentum (e.g., a blood flow). Although demanding, a motor of this type can be devised copying some of the chemiomechanics of proteins such as kinesin, myosin, or other motor proteins. We will not even attempt providing a realistic picture of the detailed mechanism of transport they implement, which is still a hot topic in biochemistry (Kolomeisky and Fisher 2007, Gennerich and Vale 2009, Hirokawa et al. 2009a, Hirokawa et al. 2009b, Wang and Schwarz 2009, Andreasson et al. 2011). Instead, let us consider a rather oversimplified model where a cargo (a large molecule, an organelle, a nano-object or a likely entity) may interact with its environment through a tether (Figure 1.6). The tether is a two-headed coiled-coil strand admitting enough conformational freedom to protrude or retract as a spring. Furthermore, it can reversibly link to a suitable, symmetric surface, its bound-detached state being controlled by external species. If the tether in its unbound state interacts with the surface through a symmetric potential, no net displacement is expected. The same holds true if the potential is asymmetrical but the rate of binding-detachment is equal. For a net motion to be observed we need to guarantee the standard two conditions for rectified Brownian motion, namely an asymmetrical potential landscape *and* non-equilibrium conditions. This may happen if the cargo is shaped in such a way to selectively hinder at least one of the two species (ADP and

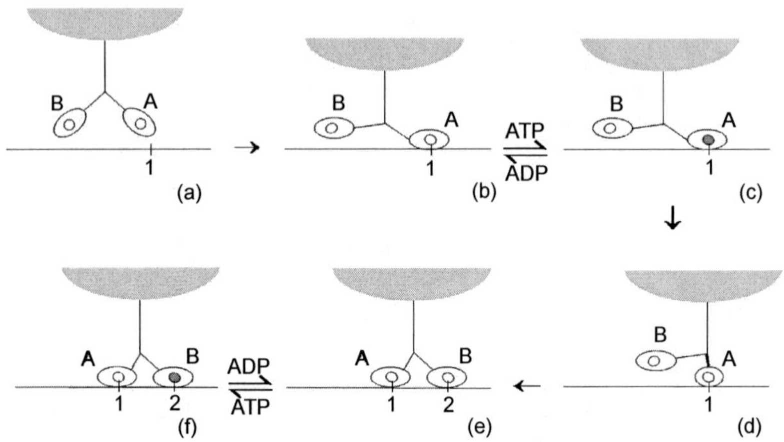

Figure 1.6: Hand-by-hand motion of a model motor protein. (a) The heads of the cargo tether, initially detached from the surface, wander until one of them (A) hits a surface site (b). An ATP molecule enters the pertinent catalytic core (c), leading to the shrinking of the tether neck (d) at the expense of ATP. This forces the other head (B) to poke sites on the right of the A-head landing site (e), binding in turn an ATP molecule (f). Note that the whole mechanism is directional only if ATP-ADP exchange preferentially occurs on one of the two cargo sides.

ATP for kinesin) ruling the attachment-detachment cycle. In greater detail, if we start with the cargo detached from the surface, the two heads of the tether, both accommodating an ADP molecule in their catalytic cores, may wander onto the surface searching for a binding site. If one of the heads encounters the surface, then it binds onto it, releases its ADP molecule and rapidly replaces it with an ATP molecule. Such a replacement determines a contraction of the molecular neck linker interposed between the two heads, the energy for the contraction being taken from the conversion of ATP into ADP. The second head can now just wander far enough to bind to a site on the right of the last occupied one—releasing in turn its ADP and linking an ATP. This triggers a new contraction of the neck linker, the whole molecule progressing a further step on the right. However, the whole mechanism properly works if and only if the conformation of the cargo is such that the ADP-ATP turnover may reach the relevant tether head with a higher rate when the head is poking sites in one direction, for example on the right. If this happens the probability that the tether gets attached on a site on the right is clearly higher than the probability it sticks on a site on the left—so that in the new bound state the head progresses one step on the right, the final result of the whole cycle being that the molecule (cargo + tether) shifts to the right. Nonetheless, the conformational hindrance due to the cargo shape is a necessary but not sufficient condition to bias the Brownian motion. Should the ADP-ATP exchange at the tether heads occur under thermodynamic control (at equilibrium), both the leftmost and the rightmost head would have the same probability to refurbish with ATP—and no net motion would be observed. It is instead the finiteness of the ATP-ADP exchange rate that enables motion. What actually keeps the system away from equilibrium is the fact that the exchange rates are different (and relatively slow). Thus, the exchange process occurs under kinetic control, and is dominated by the diffusive step letting ATP travel from the solution to the pertinent catalytic core.

The lesson one can learn from this (oversimplified, yet physically appropriate) model is that motors can be added to nanobots even when no asymmetric potential landscapes are provided by the environment. The kinesin-like drifting trick (often referred to as hand-to-hand motion mechanism) can also be implemented in artificial devices. Let a nanobot climb a capillary, and let it have tethers able to bind to the capillary walls with a mechanism similar to that expressed by kinesin. Should the nanobot have an asymmetric shape impeding the diffusion of ATP toward the tether heads in one direction, it could be instructed to run toward a target molecule simply by orienting its non impeding (smaller) side toward it—for instance using the expedient of our second toy model. In the absence of a significant target gradient its Brownian motor would clearly remain ineffective, the automaton blindly moving along with the blood flow. However, upon perception of a target molecule gradient the nanobot would activate its motor, stepping away from the blood flow and slithering onto the capillary wall up to its target.

SUMMARY

In this chapter we have discussed some prospectives—of physically yet plausible scenarios—of Nanomedicine. The possibility of devising, *using the technology available these days*, swarms of nanobots patrolling a living organism and eventually acting as a therapeutic agent toward malignant cells are described. We have also discussed in some detail the several issues of motility *in vivo* and at the nanoscale, showing how, further to having nanobots drifting along with blood in the circulatory system, additional maneuvering engines may be implemented using the art of rectified Brownian motors. One important point stressed is that Brownian motors are not simply propelling agents but also have the capability of driving nano-objects toward a target. This coupled ability of moving and directing objects has been shown to be rather common in biology, and is possibly the reason of the high efficiency of a host of biomolecules acting *in vivo*. The lesson from Nature (bio)physics is learning and appears to be usable in several classes of nano-objects, adding a special spin in the case of Nanomedicine. The potential of patrolling nanobots directed toward classes of cells instead of sampling them stochastically may significantly raise their efficiency, also reducing the invasiveness of either resident nanobots or on-purpose administrable nanodrugs.

For the benefit of the still skeptical reader we may add that we do not expect that the simple feasibility of a technology translate into its deployment. It is quite obvious that, beyond technology and science, two additional ingredients are required. One is economical, as the involvement of microelectronic fabs into such an explorative area does require a market to develop. While it is easy to speculate that, in its maturity stage, the Nanomedicine market will lead by orders of magnitude that of transistors, it is nonetheless quite obvious that the hill to overcome and to get to that stage is truly eminent, and requires diverting high-cost resources to a high-risk sector. Opening appropriate niches of market to motivate investments seems to be needed. Singling out medical issues wherever no standard diagnostic approach looks satisfactory might provide the due impulse to a first local deployment of this approach. We mentioned in the Introduction the well known case of pancreatic cancer, still one of the most deadly tumors in the US (and seemingly in the world), where the lack of reliable markers enabling early detection is the main cause of the high mortality associated with such a disease. Physicians could also assess other pathologies with likely characteristics. Explorative use of an even simplified version of this approach would utilize active nano-object parts of the paradigm of critical pathology care, letting nanobots step into the pharmaceutical arena.

A second, possibly not so minor issue to be dealt with is ethics. Inoculation in a living organism of swarms of nano-objects with a limited, yet non-null, intelligence will probably raise some initial concerns—probably not just in the patients—as it happened during the first campaigns of mass vaccinations in the early 19th century. Side technological exploitation (not just testing) on animals might provide a path to acquaint the large public with the idea, although in this case it would be much more difficult to single out an animal pathology relevant enough to motivate the fabrication of even simplified nano-objects.

In conclusion, one might observe , as often happens, that scientific and technological issues are the simplest part in making of a new deployable technology. Notwithstanding, the rapid growth of interest even in the medical community toward nanotechnology has led to the clinical use of nanodrugs so quickly that the second release of Nanomedicine tools, expressing structured and functioning nano-objects, will possibly find its way to experimentation in spite of the many difficulties we can imagine and the hills to be overcome—as the view, beyond the hills, seems so promising and exciting and is certainly worth giving a glance!

References

An, Z. and H. Möhwald and J. Li. 2006. pH controlled permeability of lipid/protein biomimetic microcapsules. Biomacromolecules 7: 580–585.

Andreasson, J.O.L. and B.E. Clancy, W.O. Hancock and S.M. Block. 2011. Mechanochemical Properties of the Kinesin-2 Motor, KIF3A/B, Studied by Optical Trapping. Biophysical Journal 100: 122.

Arbuckle, D. and A.A.G. Requicha. 2004. Active self-assembly. *In:* Proc. IEEE Intl. Conf. on Robotics and Automation—ICRA 04. New Orleans, USA.

Astumian, R.D. 1997. Thermodynamics and kinetics of a Brownian motor. Science 276: 917.

Astumian, R.D. and I. Derényi. 1998. Fluctuation driven transport and models of molecular motors and pumps. European Biophysics Journal 27: 474–489.

Berlin, J.M. and A.D. Leonard, T.T. Pham, D. Sano, D.C. Marcano, S.Y. Yan, S. Fiorentino, Z.L. Milas, D.V. Kosynkin, B.K. Price, R.M. Lucente-Schultz, X.X. Wen, M.G. Raso, S.L. Craig, H.T. Tran, J.N. Myers and J.M. Tour. 2010. Effective Drug Delivery, *In Vitro* and *In Vivo*, by Carbon-Based Nanovectors Noncovalently Loaded with Unmodified Paclitaxel. ACS Nano 4: 4621–4636.

Bonabeau, E. and M. Dorigo and G. Theraulaz. 1999. Swarm intelligence: From natural to artificial systems. Oxford University Press, New York, USA.

Brown, J.M. 2000. Exploiting the hypoxic cancer cell: mechanisms and therapeutic strategies. Molecular Medicine Today 6: 157–162.

Cashion, M.P. and T.E. Long. 2009. Biomimetic design and performance of polymerizable lipids. Accounts of Chemical Research 42: 1016–1025.

Cavalcanti, A. and R.A. Freitas Jr. 2005. Nanorobotics control design: A collective behavior approach for medicine. IEEE Transactions on NanoBioscience 4: 133–140.

Cavalcanti, A. and T. Hogg, B. Shirinzadeh and H.C. Liaw. 2006. Nanorobot communication techniques: A comprehensive tutorial. pp. 1–6. *In:* ICARCV '06. 9th International Conference on Control, Automation, Robotics and Vision, Singapore, IEEE.

Cerofolini, G.F. and V. Casuscelli, A. Cimmino, A. Di Matteo, V. Di Palma, D. Mascolo, E. Romanelli, M.V. Volpe and E. Romano. 2007. Steps farther towards micro-nano-mole integration via the multispacer patterning technique. Semiconductor Science and Technology 22: 1053–1060.

Cerofolini, G.F. and P. Amato and E. Romano. 2008. The multi-spacer patterning technique: a non-lithographic technique for terascale integration. Semiconductor Science and Technology 23: 8.

Cerofolini, G. and D. Narducci and E. Romano. Litho-to-Nano Link. pp. 1890–1900. *In:* J. A. Schwarz, C. I. Contescu and K. Putyera, [eds.] 2009. Dekker Encyclopedia of Nanoscience and Nanotechnology. 2nd edn. Taylor & Francis, London, UK.

Cerofolini, G. and P. Amato, M. Masserini and G. Mauri. 2010. A Surveillance System for Early-Stage Diagnosis of Endogenous Diseases by Swarms of Nanobots. Advanced Science Letters 3: 345–352.

Choi, H.S. and W. Liu, P. Misra, E. Tanaka, J.P. Zimmer, B.I. Ipe, M.G. Bawendi and J.V. Frangioni. 2007. Renal clearance of quantum dots. Nature Biotechnology 25: 1165–1170.

Cui, Y. and Q. Wei, H. Park and C.M. Lieber. 2001. Nanowire Nanosensors for Highly Sensitive and Selective Detection of Biological and Chemical Species. Science 293: 1289–1292.

Eichhorn, R. and P. Reimann, B. Cleuren and C. Van den Broeck. 2005. Moving backward noisily. Chaos: An Interdisciplinary Journal of Nonlinear Science 15: 026113–9.

Ferri, M. and F. Suriano, A. Roncaglia, S. Solmi, G.F. Cerofolini, E. Romano and D. Narducci. 2011. Ultradense silicon nanowire arrays produced via top-down planar technology. Microelectronic Engineering 88: 877–881.

Feynman, R.P. and R.B. Leighton and M. Sands. 1966. The Feynman Lectures on Physics. Addinson-Wesley Reading, MA, USA.

Fox, R.F. 1998. Rectified Brownian movement in molecular and cell biology. Physical Review E 57: 2177.

Freitas, R.A. 2005. Current status of nanomedicine and medical nanorobotics. Journal of Computational and Theoretical Nanoscience 2: 1–25.

Freitas, R.A. 2006. Pharmacytes: An ideal vehicle for targeted drug delivery. Journal of Nanoscience and Nanotechnology, 6 9: 2769–2775.

Gennerich, A. and R.D. Vale 2009. Walking the walk: how kinesin and dynein coordinate their steps. Current Opinion in Cell Biology 21: 59–67.

Hahm, J. and C.M. Lieber. 2004. Direct ultrasensitive electrical detection of DNA and DNA sequence variations using nanowire nanosensors. Nano Letters 4: 51–54.

Hirokawa, N. and R. Nitta and Y. Okada. 2009a. The mechanisms of kinesin motor motility: lessons from the monomeric motor KIF1A. Nature Reviews Molecular Cell Biology 10: 877–884.

Hirokawa, N. and Y. Noda, Y. Tanaka and S. Niwa. 2009b. Kinesin superfamily motor proteins and intracellular transport. Nature Reviews Molecular Cell Biology 10: 682–696.

Hogg, T. 2008. Distributed control of microscopic robots in biomedical applications. Advances in Applied Self-organizing Systems 147–174.

Jeon, Y.B. and R. Sood, J. Jeong and S.G. Kim. 2005. MEMS power generator with transverse mode thin film PZT. Sensors and Actuators A: Physical 122: 16–22.

Kolomeisky, A.B. and M.E. Fisher. 2007. Molecular motors: a theorist's perspective. Annual Review of Physical Chemistry 58: 675–695.

Li, Z. and Y. Chen, X. Li, T.I. Kamins, K. Nauka and R.S. Williams. 2004. Sequence-specific label-free DNA sensors based on silicon nanowires. Nano Letters 4: 245–247.

Loncar, M. and C.J. Barrelet, B. Jiming, H.G. Park, C.M. Lieber and F. Capasso. 2006. Semiconductor nanowires embedded in optical microcavities. *In:* 2006 Conference on Lasers & Electro-Optics/Quantum Electronics and Laser Science Conference (CLEO/QELS), IEEE.

Mahmud, G. and C.J. Campbell, K.J.M. Bishop, Y.A. Komarova, O. Chaga, S. Soh, S. Huda, K. Kandere-Grzybowska and B.A. Grzybowski. 2009. Directing cell motions on micropatterned ratchets. Nature Physics 5: 606–612.

Morigaki, K. and T. Baumgart, U. Jonas, A. Offenhäusser and W. Knoll. 2002. Photopolymerization of diacetylene lipid bilayers and its application to the construction of micropatterned biomimetic membranes. Langmuir 18: 4082–4089.

Nagrath, S. and L.V. Sequist, S. Maheswaran, D.W. Bell, D. Irimia, L. Ulkus, M.R. Smith, E.L. Kwak, S. Digumarthy, A. Muzikansky, P. Ryan, U.J. Balis, R.G. Tompkins, D.A. Haber and M. Toner. 2007. Isolation of rare circulating tumour cells in cancer patients by microchip technology. Nature 450: 1235–1239.

Narducci, D. 2011. Biosensing at the Nanoscale: There's Plenty of Room Inside. Science of Advanced Materials 3: 426–435.

O'Farrell, N. and A. Houlton and B.R. Horrocks. 2006. Silicon nanoparticles: applications in cell biology and medicine. International Journal of Nanomedicine 1: 451–472.

Pankhurst, Q.A. and J. Connolly, S.K. Jones and J. Dobson. 2003. Applications of magnetic nanoparticles in biomedicine. Journal of Physics D—Applied Physics 36: R167–R181.

Patolsky, F. and G. Zheng and C.M. Lieber. 2006a. Nanowire sensors for medicine and the life sciences. Nanomedicine 1: 51–65.

Patolsky, F. and G.F. Zheng and C.M. Lieber. 2006b. Fabrication of silicon nanowire devices for ultrasensitive, label-free, real-time detection of biological and chemical species. Nature Protocols 1: 1711–1724.

Prigogine, I. and R. Lefever. 2007. Coupling between Diffusion and Chemical Reactions, in Advances in Chemical Pysics, vol. 39, John Wiley & Sons, New York (USA).

Requicha, A.A.G. 2003. Nanorobots, NEMS, and nanoassembly. Proceedings of the IEEE 91: 1922–1933.

Sherwood, L.M. and E.E. Parris and J. Folkman. 1971. Tumor Angiogenesis: Therapeutic Implications. New England Journal of Medicine 285: 1182–1186.

Soppimath, K.S. and T.M. Aminabhavi, A.R. Kulkarni and W.E. Rudzinski. 2001. Biodegradable polymeric nanoparticles as drug delivery devices. Journal of Controlled Release 70: 1–20.

Stern, E. and J.F. Klemic, D.A. Routenberg, P.N. Wyrembak, D.B. Turner-Evans, A.D. Hamilton, D.A. LaVan, T.M. Fahmy and M.A. Reed. 2007. Label-free immunodetection with CMOS-compatible semiconducting nanowires. Nature 445: 519–522.

Wagner, V. and A. Dullaart, A-K. Bock and A. Zweck. 2006. The emerging nanomedicine landscape. Nature Biotechnology 24: 1211–1217.

Wang, X. and T.L. Schwarz. 2009. The mechanism of Ca^{2+}-dependent regulation of kinesin-mediated mitochondrial motility. Cell 136: 163–174.

Weinstein, S. and D. Peer. 2010. RNAi nanomedicines: challenges and opportunities within the immune system. Nanotechnology 21: 232001.

Welter, M. and H. Rieger. 2010. Physical determinants of vascular network remodeling during tumor growth. The European Physical Journal E: Soft Matter and Biological Physics 33: 149–163.

Williams, C.B. and R.B. Yates. 1996. Analysis of a micro-electric generator for microsystems. Sensors and Actuators A: Physical 52: 8–11.

Zheng, G.F. and F. Patolsky, Y. Cui, W.U. Wang and C.M. Lieber. 2005. Multiplexed electrical detection of cancer markers with nanowire sensor arrays. Nature Biotechnology 23: 1294–1301.

Zhou, X.T. and J.Q. Hu, C.P. Li, D.D.D. Ma, C.S. Lee and S.T. Lee. 2003. Silicon nanowires as chemical sensors. Chemical Physics Letters 369: 220–224.

2

Biotechnology-Utilized Nanopore for Single-Molecule Investigation

Ji Wook Shim[1,*] and *Li-Qun Gu*[2,*]

INTRODUCTION

Principle of Nanopores

The nanopore sheds light with its potential biotechnology applications, and moreover, is being intensively investigated and developed as the next-generation of DNA-sequencing technology with US$1,000 genome sequencing program supported by the US National Human Genome Research Institute (NHGRI). This chapter will introduce our recent research on biological and synthetic nanopores, new methods of investigating single-molecule kinetics with nanopore, and nanopore-based sensing devices.

The nanopores in dimension of single molecules are emerging as a powerful research and sensing technique with broad potential applications in biotechnology and biomedical engineering. There are biological nanopores and synthetic nanopores, both being intensively and successfully investigated. Biological nanopores, such as transmembrane protein pores or ion channels, can open and close in accordance of chemical, electrical, and mechanical stimuli (Hille 2001). In the open state, these bio-pores selectively transport ions across the cell membrane

[1]Department of Electrical Engineering, Stinson-Remick Hall, University of Notre Dame, Notre Dame, IN 46556; E-mail: Jiwook.Shim.5@nd.edu

[2]Department of Biological Engineering and Dalton Cardiovascular Research Center, University of Missouri, Columbia, MO 65201; E-mail: gul@missouri.edu

*Corresponding author

with an applied transmembrane voltage, and the generated pico-Ampere range ionic current through the pore can be accurately measured. Because the correlation between the open state probability and the stimulus intensity such as ligand concentration can be established through protein engineering (Hille 2001), these engineered protein pores can work as a biosensor for detecting target specific reactions (Bayley and Cremer 2001). Different from the self-assembled biological nanopore, synthetic nanopore is a tiny hole drilled artificially on the solid-state membrane such as glass, Si, SiO2, Si3N4, Al_2O_3 and Graphene. The nanopore is mainly punched by etching or electron beams at SEM or TEM. They can also, they can be engineered for specific target sensing through chemical modification on the nanopore interior surface.

The nanopore is distinct from other sensing devices. It works in the manner of detecting an individual molecule that binds within the pore. The advanced single-molecule sensing mechanism makes the nanopore more versatile because it can recognize and quantify a specific target molecule among mixed groups. The principle of nanopore single molecule detection is shown in Fig. 2.1. The nanopore is clamped and sealed by two apparatuses with high resistance (>10GΩ). The pore itself forms the only pathway that allows the target samples to migrate between two reservoirs. The specific formation of the biological nanopore and synthetic nanopore

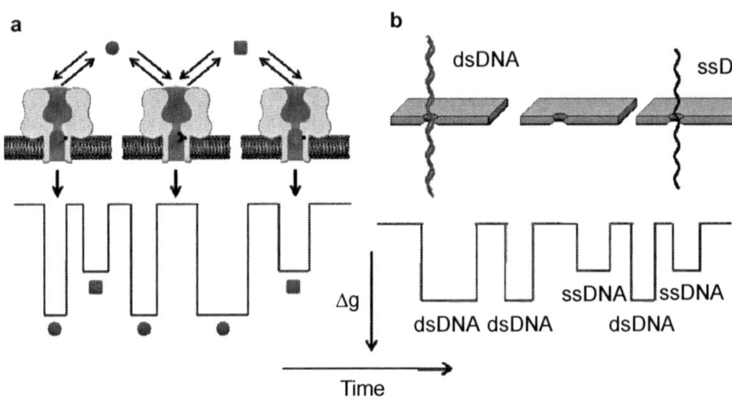

Figure 2.1: Cross-sectional view of single-molecule detection in protein nanopore and synthetic nanopore. (a) Different targets (shown as blue ball and red rectangle) bind to the engineered protein pore competitively and reversibly. Each target produces its own electrical signature with a specific current blocking amplitude and duration. (b) Synthetic nanopore with a double-stranded and single-stranded DNA threading through. This process produces different current blocking profile according to the size of threading molecules. Through single-molecule detection in the nanopore, targets can be identified from their electrical signature and quantified from the binding frequency.

Color image of this figure appears in the color plate section at the end of the book.

are completely different. For the synthetic nanopore, the solid-state membrane is sealed with flexible hydrophobic materials such as PDMS (polydimethylsiloxane) and silicone O-ring. Only the very tiny area of the membrane where a nanopore is drilled is hung on the substrate, while the rest of the area is supported with robust substrate such as silicon, and large enough so that it is suitable to seal with commercially available sealers. Finally the complex is clamped by two apparatuses. By comparison, the protein nanopores should be incorporated into a membrane such as lipid bilayer in order to perform its functions. Due to the fragility of artificially formed lipid bilayer, a hydrophobic film is used as a support to hold the lipid bilayer and isolate the solutions on two sides of the bilayer. Briefly, a hydrophobic film such as Teflon is clamped by two apparatuses as a partition that isolates the solution into cis and trans. A planar lipid bilayer is formed over a tiny ~100 μm puncture at the center of the partition film. The protein pore is then introduced to one side of recording cell, which spontaneously incorporates into the lipid bilayer to form the only passage between two recording cells as an ionic flow path. Both sides of the recording cell are connected with a pair of Ag/AgCl electrodes, from which the voltages across the pore is applied and the nanopore pico-Ampere current is recorded.

The nanopore allows a certain level of ionic current at an applied voltage. The ionic current can be reduced when a single molecule binds to the pore. The ionic current resumes to the original level as the bound molecule is released from the pore. The change in conductance, Δg, made by each molecule of the same target are identical, and thus acts as an electrical signature for target identification. The binding duration of single molecules is stochastic following the exponential distribution. The fitting of the distribution gives the average binding duration, τ_{off}, which returns the dissociation rate constant k_{off} by $k_{off} = 1/\tau_{off}$. The occurring frequency of the target binding events can be obtained by measuring the interval between adjacent events, which is also stochastic and forms the exponential distribution. The fitted constant, τ_{on}, can be transformed to the event frequency by $f = 1/\tau_{on}$. As the frequency is target concentration-dependent, i.e., $f = k_{on}$ [T], k_{on} where the association rate is constant, one can use this expression to quantify the target.

Biological Nanopore-based Single Molecule Detection

Among biological nanopore, α-hemolysin (αHL) has been intensively investigated and developed for various nanopore-based biotechnology applications. αHL is an exotoxin secreted by the bacterium Staphylococcus aureus. The monomer of αHL contains 293 amino acids, and it can be self-assembled into a heptameric protein pore in the lipid bilayer membrane.

This 10-nm long protein pore consists of a 4.6-nm wide mushroom-shaped nanocavity on the top, a 2-nm wide β-barrel below of the nanocavity, and a 1.4 nm constriction between the nanocavity and β-barrel (Song et al. 1996) (Fig. 2.2).

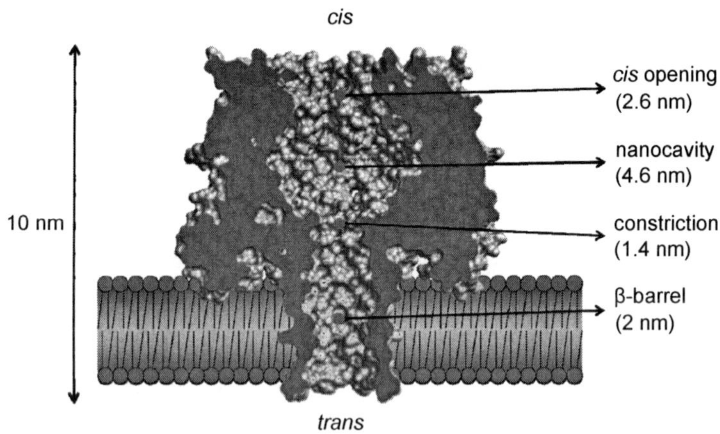

Figure 2.2: Molecular diagram of the α-hemolysin heptameric protein pore in the lipid bilayer (Song et al. 1996). The dimensions at specific sites are provided.

Color image of this figure appears in the color plate section at the end of the book.

Bayley's group have pioneered the applications of biological nanopore in biosensing (Bayley and Cremer 2001, Bayley and Jayasinghe 2004). αHL has been engineered with different chemical properties for various detection purposes. For example, the pore engineered with a tetra-histidine motif in one of seven subunits at the β-barrel can reversibly capture a single divalent ion. The signatures can distinguish different ions such as Zn^{2+} and Co^{2+} in the mixture at nano-molar concentration (Braha et al. 2000). By socketing double arginine rings around the constriction site, the αHL pore became capable of detecting different phosphate compounds, and discriminating the second messenger inositol triphosphate (IP3) (Cheley et al. 2002). The larger protein molecules that cannot enter the pore can also be detected through a probe-terminated linker attached in the lumen (Movileanu et al. 2000). In a programmable approach, a ring-shaped molecular adapter can be lodged at the constriction site to narrow down the pore size, therefore allowing better discrimination of drug molecules in similar structures (Gu et al. 1999) and chiral enantiomers (Kang et al. 2006).

αHL has also been investigated as a potential next-generation DNA sequencing technique. When a single-stranded DNA (ssDNA) translocates through the pore driven by the voltage, it occupies a portion of the nanopore, resulting in partial current blockade (Kasianowicz et al. 1996).

Theoretically, each DNA base could be discriminated if its translocation through the pore generates a specific current change. This nanopore-based single DNA sequencing is supposed to be simpler, more rapid and cost-efficient. Compared with other sequencing technologies, the nanopore sequencing method is label-free, enzyme-free, ling, and does not need DNA amplification. Towards this goal, a wealth of information and knowledge has been acquired on DNA and RNA translocation through nanopores. These studies allow us to understand how the specific sequence modulates the current blocking in terms of amplitude and duration depending on environmental conditions such as temperature, applied voltage, and chemical modification (Kasianowicz et al. 1996, Akeson et al. 1999, Meller et al. 2000, Meller et al. 2001, Deamer and Branton 2002, Branton et al. 2008, Mitchell and Howorka 2008, Ding et al. 2009, Gao et al. 2009).

Recently, new strategies for nanopore-based DNA base-discrimination have been proposed. For example, a ssDNA in the form of complex linked with large molecule such as DNA/RNA hairpin (Ashkenasy et al. 2005) or Strepavidin (Stoddart et al. 2009) at the end of ssDNA can be immobilized in the nanopore. The ssDNA, therefore, stays in the pore a long time instead of translocating. The current blocking difference allows distinguishing all four DNA bases A, G, T, and C with immobilized DNAs in the pore lumen (Stoddart et al. 2009). As an advanced sequencing concept in terms of single-molecule, a novel DNA sequencing strategy in nanopore has been implemented. In this strategy, the exonuclease is used to digest a DNA into single nucleotides, then when these digested nucleotides sequentially translocate through an adapter-attached pore, they can be accurately recognized by the adapter (Clarke et al. 2009). This scaffold toward nanopore-based DNA sequencing with high fidelity can discriminate between all four bases.

In addition to DNA detection, the nanopore is also a versatile single-molecule instrument to investigate various molecular processes. The nanopore very sensitively reacts to the change in conformation, dimension and orientation of a molecule in the pore, so that the conductance signature reveals the state of single molecules. αHL has been used to identify intermediate states of a chemical reaction, to uncover pathways, and to quantify the kinetics of single-molecule by being equipped with a reactant in the lumen (Luchian et al. 2003). Nanopores are also used to examine the unzipping process of a single DNA or RNA molecule (Kasianowicz et al. 1996, Vercoutere et al. 2001, Mathe et al. 2004, Nakane et al. 2004, Ashkenasy et al. 2005, Astier et al. 2006). dsDNA or dsRNA with an overhang can be trapped from the cis side of the pore. The single-stranded part can enter the β-barrel while the double-stranded region is anchored in the nanocavity. Eventually, the single-stranded DNA region can be pulled by the electric field, resulting in unzipping the double-stranded DNA region (Vercoutere et al. 2001, Mathe et al. 2004, Nakane et al. 2004, Ashkenasy et al. 2005, Astier et al. 2006).

Nanopores have been successfully used to measure the binding of enzymes to their DNA substrates. When the enzyme/DNA complex is introduced to the pore, the DNA enters the pore while the enzyme is stopped from entering due to size restriction. The DNA in the pore pulls over, and eventually, detaches itself from the enzyme protein. This principle mimics a molecular force microscope that can examine the force involved in interaction between the enzyme and DNA target at single molecule resolution (Akeson et al. 2007, Tropini and Marziali 2007).

There are other protein nanopores that are being explored for sensing applications. Engineered peptide gramicidin can form a channel in tethered lipid bilayer membrane for protein detection (Cornell et al. 1997), and modified gramicidin for chemical reaction detection (Capone et al. 2007). Another protein pore formed by the porin MspA of *Mycobacterium smegmatis*, has recently presented the ability of detecting DNA translocation (Butler et al. 2008) and discriminating DNA bases with ssDNA attached to hairpin-duplex while it stays longer in the pore until the hairpin-duplex dissociates (Derrington et al. 2010).

Solid-state Nanopore-based Single-molecule Detection

With the development of micro- and nano-technologies, it becomes possible to fabricate devices in a very tiny scale. It has especially been able to drill a pore in single molecular size on a solid-state membrane with the state-of-art technique. Researchers were able to drill a tiny pore on glass, Si, SiO_2, Si_3N_4, and graphene. There was nano-meter scale pore on Si_3N_4 through the etching technique (Schmidt et al. 2000), and on glass by the ion-track etching technique (Fertig et al. 2001). The track etching technique was used to generate 15 nm pore on polymer foil of polyethylene terephthalate (PET) (Siwy and Fulinski 2002). Ion beam sculpting method has been used to form nano-meter scale pore by exposing Ar^+ ion beam on the back side Si3N4 on which a cavity has been made by Focused Ion Beam (FIB) (Li et al. 2001). There was an indirect method to produce smaller nanopores by shrinking larger nanopores. It used high-intensity wide-field illumination with electron and researchers found that it shrinks the pore on SiO_2 (Storm et al. 2003). Direct nano-meter pore forming method through TEM was also well-developed. The nanopore was drilled by very focused electrons at TEM on SiN, Al_2O_3, and Graphene (Heng et al. 2004, Zandbergen et al. 2005, Krapf et al. 2006, Garaj et al. 2010, Merchant et al. 2010, Schneider et al. 2010, Venkatesan et al. 2010).

Most research working with synthetic nanopore have characterized the translocation of single molecules, particularly DNA molecules, through the pore (Howorka and Siwy 2009). With a pore size similar to the width of DNA molecules, both linear form DNA (Storm et al. 2005) and the DNA with a folded structure (Li et al. 2003, Storm et al. 2005) can generate characteristic current blocks when translocating the pore. It has also been found that dsDNA can pass through a slightly narrower pore than its size, and the voltage threshold for dsDNA translocation is attributed to the stretch of dsDNA (Heng et al. 2005).

In addition, synthetic nanopore can also be utilized for investigating relatively big protein molecules (Siwy et al. 2005, Han et al. 2006). As the synthetic nanopore has been well developed in a short time, it develops various potential applications with the advantage of robust stability and controllable size in molecular dimension. It is expected that it would be beneficial to the understanding of single-molecule behavior and interactions.

Comparison between Biological-pore and Synthetic-pore

There is a well-explained comprehensive comparison between biological nanopores and artificial nanopores along with their perspectives in recent review articles (Martin and Siwy 2007, Howorka and Siwy 2009). The biggest advantage of biological protein nanopores is consistency in size and modification. Biological nanopores are created and found in nature. They are extracted and reconstituted to the artificial lipid membrane so they are in the same size with each protein pore by protein pore. They can be genetically engineered and chemically modified to have structure-directed specific receptor and/or to have site-directed specific polar characteristic (Bayley and Jayasinghe 2004). These advantages make biological nanopore selective for specific target and highly reproducible for detection.

The synthetic nanopores have superior stability in mechanical and electrical shocks, changeable and controllable pore sizes, and adaptability into array platforms. Due to these characteristics of the synthetic nanopore, they have been used for application in biotechnology and life sciences research (Saleh and Sohn 2003, Uram et al. 2006, Sexton et al. 2007, McNally et al. 2008). They function properly at high voltages without a stabilizing supporter which is required for the biological nanopore. The controllable pore size makes them versatile and suitable for the detection of different dimension target molecules. These are fabricated on diverse materials such as Si, SiO2, Si3N4, graphene, and glass, and could also be modified on surface.

Biological nanopores, inspite of their advantages of being reproducible and modifiable, face limitations such as fixed pore size and are vulnerable to mechanical and electrical disturbance. The lipid bilayer, in which protein pore is incorporated, is very essential for the biological nanopore system, but it limits the ability of the biological nanopore. Lipid membrane is easily influenced by changes in the surrounding environment. In contrast, the synthetic nanopore has demonstrated high stability against mechanical and electrical stress, and controllable pore width and length. However, it is necessary to improve superfine-reproducibility of the size of the synthetic nanopore to make it more reliable, and to explore site-targeted surface modification on the synthetic nanopore to improve versatility.

INVESTIGATION OF SINGLE-MOLECULE KINETICS THROUGH NANOPORE/NANOCAVITY

Trapping a Single-molecule in the Nanocavity of αHL

We have first utilized the nanocavity enclosed by the αHL pore to trap a single-molecule non-covalently, and demonstrated a versatile nanopore-based system for understanding ion-regulated folding and unfolding of a single G-quadruplex molecule (Shim and Gu 2008, Shim et al. 2009). G-quadruplex is guanine-rich ssDNA or ssRNA formable into intra- or intermolecular four-stranded complex (Sen and Gilbert 1988, Hardin et al. 2000, Burge et al. 2006). G-quadruplexes are also found in human telomeres, and are associated with an important role in gene regulation (Arthanari and Bolton 2001). They are found especially at the cancer cell and, therefore, they are target drugs for cancer treatment (Neidle and Parkinson 2002). Many synthetic quadruplexes have been designed with high affinity to certain proteins, and thus can be used as biosensors (Bock et al. 1992) and potent pharmaceuticals (Tang et al. 2007). In addition, synthetic quadruplexes can be used as exceptionally tiny bricks of nano-structures (Davis and Spada 2007) and parts of nanomachines (Li and Tan 2002, Alberti and Mergny 2003).

Spontaneous folding and unfolding of a single-molecule using the nanopore has also been investigated. The target was thrombin-binding aptamer (TBA) (Shim and Gu 2008, Shim et al. 2009). TBA is a short oligonucleotide of 15 bases (GGTTGGTGTGGTTGG) folding into a quadruplex with two tetrads sandwiching one cation (Marathias and Bolton 2000). When folded, TBA can bind to its target thrombin with high affinity, and thus can be used as an inhibitor to the thrombin clotting activity (Bock et al. 1992), and as a biosensor for thrombin detection (Heyduk and Heyduk 2002).

Initially we had thought that the nanocavity in αHL can trap a single TBA G-quadruplex. There are two identification characteristics of electrical current signatures when TBA is added to the cis side of the pore (Shim and Gu 2008, Shim et al. 2009) (Fig. 2.3). One is a full block with a short-lived duration (~100 μs) and the other is a partial block with a long-lived duration (~10 s). The short-lived events are generated by the linear form (unfolded) TBA translocating through the pore, while the long-lived events are attributed to the trapping of a single TBA in form of G-quadruplex (folded) in the nanocavity domain. The difference in the two events can be explained by the size of TBA in folded and unfolded forms. When TBA is in the linear form, it can easily pass through the constriction site (~1.4 nm) and β-barrel (~2 nm), generating spike-like translocation events. However, TBA in the G-quadruplex form (~ 2 nm) is smaller than cis opening but

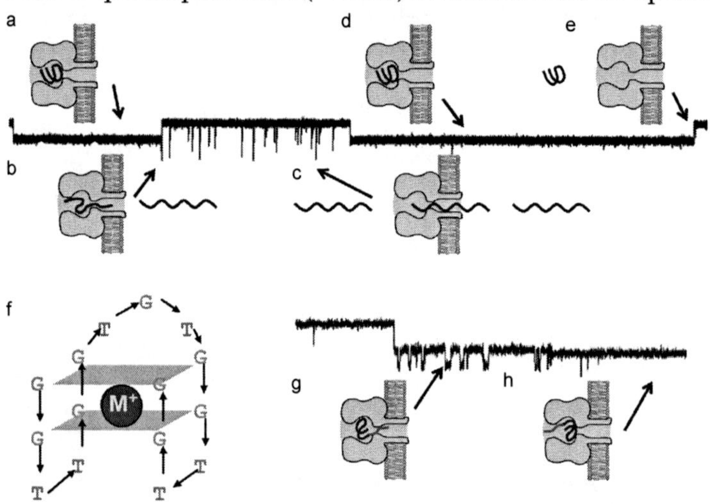

Figure 2.3: Investigation of folding and unfolding of a single G-quadruplex through a nanocavity enclosed in α-hemolysin. (a) The long-lived shallow current blocking indicates capturing a folded G-qaudruplex in the nanocavity. (b) G-quadruplex unfolds in the nanocavity and leaves the protein pore to the other side. This translocation event of unfolded aptamer is identified by terminal-spike. (c) Unfolded aptamers in linear form translocate through the protein pore. (d) Another folded G-quadruplex enters into the nanocavity. (e.) G-quadruplex escapes from the nanocavity to the entrance. (f) Schematic graph of folded Thrombin-Binding Aptamer G-quadruplex. It has two G-tetrads formed by 1, 6, 10, and 15 for top tetrad, and 2, 5, 11, and 14 for bottom one. A monovalent cation intervenes between two tetrads. (g) Current characteristic of tagged TBA. It produces an additional current blocking which indicates that the G-quadruplex is vibrating in the nanocavity. (h) The disappeared additional current blocking is shown by tag-TBA heading to the other side, which indicates that the G-quadruplex rotates in the nanocavity as well.

(Note: The data traces have been modified from its original. Spike-noise is hidden by figure a and b. Data trace for g and h are obtained in different time region of the same data trace but merged to show two events together.)

Color image of this figure appears in the color plate section at the end of the book.

larger than β-barrel, thus it can be trapped in the nanocavity, but stopped at the constriction site from entering the β-barrel. This configuration allows a large ionic current through the unoccupied portion of the nanocavity and β-barrel, generating long partial blocks. In addition, including thrombin, which is larger than the pore size (Padmanabhan et al. 1993), can remove all these long partial blocks, confirming that thrombin has captured all the G-quadruplex TBA, leaving no free G-quadruplex that can be trapped in the pore. An interesting observation is that the long partial block is terminated with a full spike-like current block. This type of block is due to the linear form DNA translocation through the pore, therefore we concluded that the trapped G-quadruplex can unfold into the linear form DNA. This unfolding was later identified as a spontaneous process because the duration of long block, or the lifetime of the G-quadruplex is independent to the applied voltage (Shim and Gu 2008). From signature blocks by a tagged-TBA G-quadruplex in the nanocavity, we found that the G-quadruplex stays and vibrates at the bottom of the nanocavity near the entrance of the barrel. In summary, the main observations from this study include: (1) the nanocavity of the nanopore can be utilized to trap a single-molecule, (2) two different TBA in forms of G-quadruplex and linear single-stranded DNA can be discriminated by an electrical current characteristic, and (3) conformational change of a single-molecule such as spontaneous unfolding can be detected by nanocavity/nanopore. Therefore, the nanocavity/nanopore system is expected to be applied as a research tool for studying single-molecule activity.

Detection of Folding/unfolding Kinetics of a Single Molecule Trapped in the Nanocavity/nanopore

The G-quadruplex formation is cation-selective. A cation is coordinated by eight carbonyls from the two neighboring G-tetrads, to stabilize the G-quadruplex. We have developed an analytical method for determining the constants of equilibrium, unfolding, and folding of G-quadruplex based on dissecting the electrical current signature in the nanopore (Shim et al. 2009). The definition of each constant is given below,

Equilibrium constant: $K_f = f_{ctrl}/f_{TBAL} - 1$, where f_{ctrl} and f_{TBAL} are the frequencies of events for the control DNA and linear form TBA translocation in the pore;

Unfolding rate constant: $k_u = 1/\tau$, where τ is average duration of long-lived events;

And folding rate constant: $k_f = K_f/k_u$, where K_f is the equilibrium constant and k_u is the unfolding constant described above.

The frequency of control DNA and TBA_L are calculated from the occurrences of short events. Their ratio (f_{TBAL}/f_{ctrl}) was found to be voltage-independent. Since the control DNA has the same length as TBA with very similar specific-sequence but disabled folding capability, it is suitable to serve as a reference DNA to calculate the equilibrium constant of TBA. τ is the duration of long G-quadruplex block, and verified to be equivalent to the life time of G-quadruplex in the nanocavity. We have utilized this method to investigate how cations regulate the folding/unfolding property of the G-quadruplex. The formability of TBA into G-quadruplex represented by K_f shows that K^+, Ba^{2+}, and NH_4^+ are favored cations over Cs^+, Na^+, and Li^+ to fold TBA into G-quadruplex. However, with strong non-specific interaction between DNA-divalent cations, divalent cations Mg^{2+}, Ca^{2+}, and Ba^{2+} allowed very little frequency of control DNA translocations through the pore (10 ~ 20 times less events than with movalent cations), and Mg^{2+} and Ca^{2+} do not induce formation G-quadruplex. With the folding/unfolding constants, we have found that the high formability of G-quadrupelx induced by K^+ is greatly contributed by the slow unfolding process. The formability of Na^+- and Li^+-induced G-quadruplex were also interesting. Both Na^+ and Li^+ show low formability into G-quadruplex with similar equilibrium constants, but they encounter completely different pathways. The Na^+-induced G-quadruplex folds and unfolds in the fastest way, while Li^+ folds and unfolds G-quadruplex in the slowest way. With the equilibrium constant, it indicated that the cations in the range of 1.3 Å and 1.5 Å (K^+, Ba^{2+}, $NH4^+$) induced the G-quadruplex at high ratio. More interestingly, the cation-selective G-quadruplex formability is correlated with the volume of the G-quadruplex, which differed according to the cation species and form more tightly with cations within 1.3 ~ 1.5 Å (Shim et al. 2009).

The findings of this study suggest that the nanocavity/nanopore is a suitable tool to investigate ion-regulated characteristics and behaviors of single-molecule oligonucleotide. The nanocavity/nanopore method is not limited to guanine-rich thrombin-binding aptamer, but it is applicable to other types of quadruplex and its variants. As possible candidates, it brings up various bio-relevant intramolecular quadruplex such as the i-motif, of which the quadruplex is formed by a cytosine-rich sequence. It could be used to understand how each specific guanine contributes to the formability to G-quadruplex when combining with quadruplex loaded with a specific nucleotide by site-directed chemically-modified substitution. In addition, the nanocavity/nanopore method is able to detect molecular recognition, and to determine the force involved in the interaction between aptamer and its target. The findings and the method of this research may also contribute in discovering new molecular species with new characteristics, to develop nano-scale construction, and to design biosensors.

NANOPORE-EQUIPPED SENSING DEVICE

There have been many attempts to develop nanopore devices as biosensors for the broad range of nanopore applications. Since the major weakness of the biological nanopore-based system is caused by fragility of lipid bilayer membrane in which the protein pore is inserted, one of the efforts is to increase the stability of the membrane. Also, high-throughput applications for screening purposes such as drugs targeting membrane protein are expected, therefore it is necessary to provide a microarray platform of lipid bilayer, in which each elemental bilayer has a single responsive pore embedded. Another effort is to develop a method that can control the insertion of a single protein pore into the lipid membrane and to miniature the device. In comparison with the biological nanopore, the synthetic-nanopore has presented superior stability. A method to build artificial-nanopores at low-cost without using state-of-art facilities was also required.

Transportable Modular Biochip Equipped with a Single Protein Pore

In order to make the nanopore-based biosensor more versatile, it was necessary to be usable at off-laboratory as a real-time ready-to-use sensor, and therefore, it is required to have a robust lipid bilayer that is durable against transportation. For increasing the stability of the lipid membrane, one method has been to reduce the size of the orifice area, over which the lipid membrane forms, thus increasing membrane stability. The diameter of the orifice could be minimized to several micrometers on silicon substrate through micro fabrication (Sackmann 1996, Fertig et al. 2002). Another strategy was to support the lipid bilayer with solid or semi-solid materials. The lipid membrane was tethered onto a solid surface covalently as solid-supported bilayer (Cornell et al. 1997, Naumann et al. 2002). There were several reports on supporting the lipid bilayer with gels. Two pre-cast gel slabs were used to support the lipid bilayer membrane, which had been formed by the painted-method[1], over both sides of the membrane for double support (Costello et al. 1998). As an improved method, UV-triggered hydrogel has been used to encapsulate the membrane (Jeon et al. 2006).

Both Gu group (Shim and Gu 2007) and Bayley group (Kang et al. 2007) have autonomously reported long-lasting transportable modular biochip equipped with a single protein pore in a solidified lipid bilayer membrane. They have shown that the biochip is useful for both sensing of biomedical agents and investigation for membrane protein. The schematic

[1]Lipids are applied on the hole by using a stick, small brush, or plastic rod. Initially-formed thick film of lipids thins out spontaneously to form a lipid bilayer membrane.

diagram, prototypes, and biosensing application are shown in Fig. 2.4. The lipid membrane with a single protein pore is supported by being wedged between two hydrogel layers, and the pore performed without any impaired function. The chip can be fabricated in reduced size to facilitate portability and modular-ability. The brief introduction on chip assembly follows (Shim and Gu 2007). First, Teflon partition with an orifice (~100um) in the center is clamped by two compartment blocks, which is made of hydrophobic material such as Teflon or PDMS. Second, a desired ionic-solution mixed with 1.5% (w/v) ultra-low gelling temperature agarose is heated and maintained at room temperature. The prepared ionic solution would remain in liquid form at room temperature. Third, a single protein pore is incorporated into the lipid bilayer membrane formed by monolayer-folding method[2] (Montal and Mueller 1972). Fourth, to solidify the agarose-mixed ionic solution, the whole chip is stored in 10°C environment until the solution turns into gel. Finally, the two compartments are sealed with openings to prevent the gel from dehydration with lids in which electrodes are fixed. Prior to loading analytes, the sample cell at the upper compartment is uncovered gently to allow the passage to the target protein pore. The hydrogel is made by mixing with agarose that is in the form of a porous structure. The 2% agarose gel has 470 nm porous structure (Pluen et al. 1999). Thus it allows molecules in various weights to pass through the gel to reach the target protein pore in the lipid membrane.

The biochip has shown high durability. The solidified lipid membrane could retain a stable high giga-ohm seal (> 10GΩ in 1M KCl) longer than a

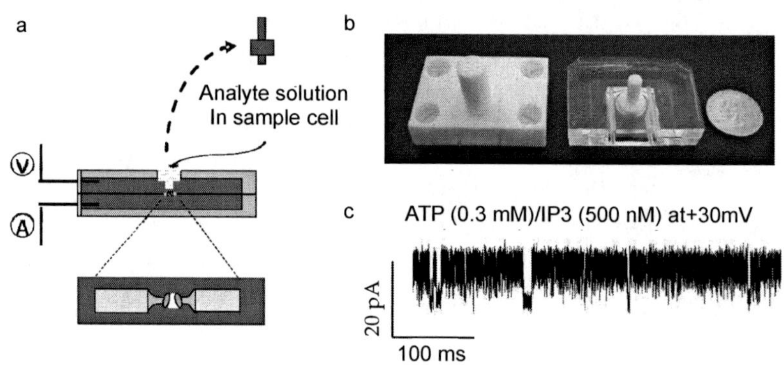

Figure 2.4: Schematic diagram, prototype and data trace of transportable modular chip. (a) Analyte can be introduced from the sample cell uncovered at the top side of the chip. Agarose layer between protein pore and the entrance of sample cell deliver the analyte.

[2]Monolayer of lipid is formed at the solution-air interface on the solution filled lower than the hole for both sides of the chamber. The bilayer is formed when the level of the solution is raised for both sides of chamber.

week by being sandwiched by hydrogel. It is a great advanced achievement for the life-time of lipid membrane when compared with several hours life-time of conventional planar lipid membrane without any support. We have used the viral potassium channel Kcv as the model pore to check that the chip made by hydrogel-sandwiched lipid bilayer is suitable for ion-channel study and does not affect channel functions. The channel functioned properly and kept its behavior through the entire length of membrane life-time. The biochip has also shown high portability. It can be reiteratively plugged on and off from the amplifier of electrophysiology recording system. Repeated connection/disconnection to the electrical device do not disturb the hydrogel-sandwiched lipid membrane and the ion channel activity in the membrane. The detached biochip can be stored independently and transported to other places where the biochip is needed for research or sensing purpose. With durability and portability while retaining ion-channel activity within lipid membrane, the biochip is facilitated to be independently storable, compatible with other systems such as a real-time ready-to-use biosensor. The biochip has been utilized with an engineered protein pore (Cheley et al. 2002) for sensing second messenger inositol 1,4,5-triphosphaste (IP3) (500 nM) in the presence of high concentration of ATP (0.3 mM).

Another featured design of the biochip is programmable. It can be equipped with any membrane protein pore for single-molecule detection, and it can be applied for researches of DNA/protein detection in genomics/proteinomics. The biochip can also be used for screening of pharmaceuticals targeting membrane proteins. It is speculated that the biochip can be used as a profitable tool for investigation such as long-term dynamics of broad range of ion-channels, which is limited to several hours at traditional ion-channel reconstitution system. Other merits of the biochip are a fast fabrication process and programmable with other materials. It can be ready-to-use in an hour by 5 min of lipid bilayer forming, 10 min of single protein pore incorporation into lipid membrane, and 10~15 of solidifying the agarose-containing solution into the gel. The lipid membrane can be supported by other polymer materials, besides agarose. Smart polymer, which reacts to specific pH or salt concentration to trigger solidification, is a good candidate of the material, since it permeates ions and molecules to target protein pore without altering the characteristic of lipid membrane and protein.

With the cutting-edge technique of micro- and nano-fabrication, the size of the biochip can be dramatically minimized so that it could improve the detection time of biochip and reduce the detection cost. About half hour is required for neutral molecule—cyclodextrin was tested for this finding—to travel through 2 mm of the porous agarose gel between the sample cell and the target protein pore (Shim and Gu 2007). However, with the minimized size of the biochip, it is expected to have travel time of the molecules decreased by at least a factor of ten through thinned out lipid membrane, and itis also, expected to reduce the amount of reagents as less as 1μL.

The pluggable biochip could be combined with other state-of-art techniques such as the micro-fluidic system for efficient sample exchange in terms of rapid, low-amount usage, and target-oriented delivery. The combination of biochip and micro-fluidic system could be achieved with the developed technique of micro-patterned hydrogels (Klajn et al. 2004, Mayer et al. 2004), and the technique of forming lipid membrane with protein pore across a micro-fluidic channel (Malmstadt et al. 2006). The biochip could have multi protein pores in the form of micro-arrays for high throughput application while each array is equipped with a single protein pore. The biochip can be fabricated in micro-array form with the edge-technique of micro-fabrication and the insertion of protein pore into each array can be achieved by a single channel assay method of quick-transferring ion channel into lipid membrane (Holden et al. 2006).

Noval Easy-made Nanopore at the Tip of Glass Capillary Pipette

We have developed a simple and accurate method of fabricating a size-controllable nanopore on the terminal of a glass capillary pipette (Gao et al. 2009) (Fig. 2.5). To fabricate the nanopore, we can melt and seal the pipette tip so that a wineglass-shaped nanocavity is formed inside the terminal. The tip is exposed to hydrofluoric acid/ammonium fluoride for external etching, and monitored by the ionic current between solutions inside and outside the pipette. A nanopore is formed once the enclosed nanocavity was perforated. The pipette tip was then transferred to an etchant-free solution to determine the pore conductance. The tip can be repeatedly etched until the desired conductance is achieved. Such a nanopore-terminated probe demonstrates several benefits. 1. The fabrication process does not require expensive technique facilities but a low-cost and simple process which is performable at any other laboratory; 2. The pore size is controllable from one nanometer to hundreds nanometers by the monitoring pore forming process through ion current measurement; 3. It has low electrical noise level; and 4. The surface of the glass pipette wall can be bio-friendly modified to be used as a biosensor. The performance of the glass-terminal nanopore has been experimentally verified for capturing single molecules and stochastic sensing. The pore size has been verified by measuring the translocation of a series of sizing molecules, such as double-stranded DNA (~2 nm), gold nanoparticles (~10 nm), and the ring-shaped cyclodextrin (~1.5 nm) (Gao et al. 2009) (Fig. 2.5). When a dsDNA translocated through the glass nanopore with a comparable size, the translocating speed can be

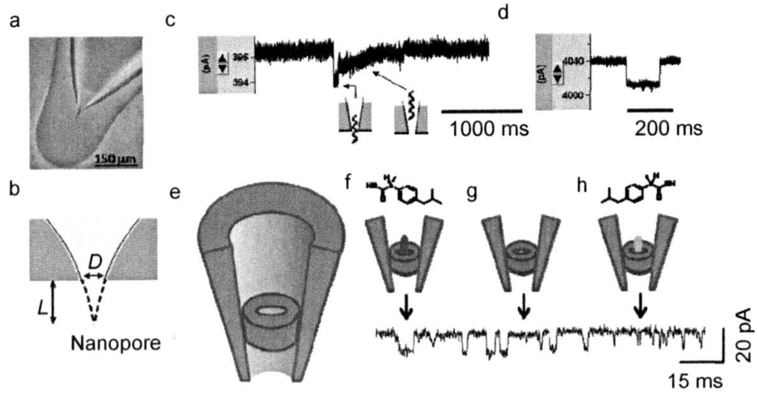

Figure 2.5: Fabrication and applications of glass-terminal nanopore. (a) The tip of a glass micropipette was melted and sealed to enclose a nanocavity at the terminal. (b) The nanopore is formed by external etching of the micropipette terminal that finally perforates the nanocavity. (c) current block generated by translocation of 1kbp dsDNA through a 2 nm pore in 1M NaCl at +100mV. With the comparable pore size to the dimension of dsDNA, the dsDNA translocation slows down and the entire translocation process can be characterized through the signature blocks. (d) Current block generated by identical dsDNA as in (c) translocation through a 7 nm pore with a much higher speed. (e) Lodging a cyclodextrin in the 1 nm glass nanopore. (f, g, and h) Single-molecule discrimination of chiral enanstomers R-ibuprofen and S-ibuprofen. When the cyclodextrin is installed in the pore, the current blocking by cyclodextrin and R-ibuprofen (f), by only cyclodextrin (g), and by cyclodextrin and S-ibuprofen at (h) can be discriminated.

slowed down, allowing monitoring translocation kinetics (Gao et al. 2009) (Fig. 2.5). Most interestingly, the fabricated 1-nm glass nanopore can trap a single cyclodextrin molecule in the pore lumen. The trapped cyclodextrin functions as an adapter since it works with biological nanopore αHL. The glass-terminal nanopore equipped with cyclodextrin becomes capable of discriminating chiral enantiomers by current blocking characteristics (Gao et al. 2009) (Fig. 2.5).

This novel glass nanopore has been developed with selectiveness for specific target molecules. We have integrated aptamers into the glass-terminal nanopore (Ding et al. 2009) (Fig. 2.6). Aptamers, also called synthesized antibodies, are *in vitro* created oligo-DNAs or RNAs (Ellington and Szostak 1990, Tuerk and Gold 1990) that selectively bind with high affinity to target ligands such as peptides (Baskerville et al. 1999), proteins (Klug et al. 1999, Wen and Gray 2004, Xiao et al. 2005), and pathogens (Jing et al. 1997, de Soultrait et al. 2002, Jeon et al. 2004). Such an aptamer-encoded nanopore can indicate obvious current change when a single target molecule binds to the immobilized aptamer in the nanopore, and detect sequential binding of a series of target molecules one by one. Thus the aptamers are excellent sensing elements for single-molecule detection

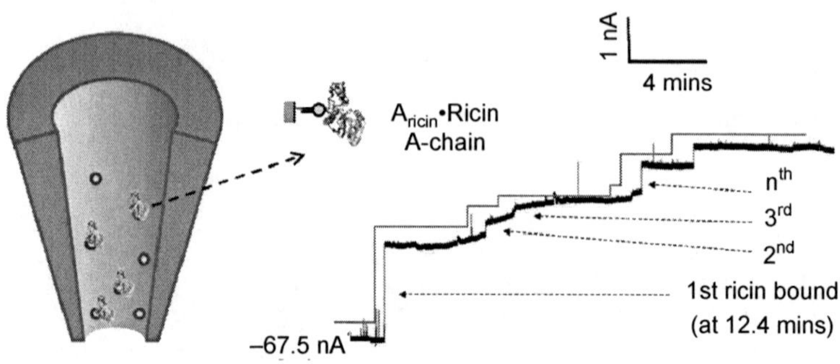

Figure 2.6: Detection of bioterrorist agent ricin with an aptamer-encoded glass-terminal nanopore. Ricin A-chain aptamers were immobilized on the internal surface of the nanopore. The aptamer-encoded nanopore was immersed in the solution containing 100 nM concentration of ricin A-chain protein and recorded at -100 mV. The stairway-like current blocking occurred as a series of molecules of ricin A-chain protein bind to the aptamer individually.

Color image of this figure appears in the color plate section at the end of the book.

in the nanopore. We have utilized the aptamer-encoded nanopore to quantitatively measure immunoglobulin E (IgE) and ricin, one of the most toxic bioterrorist agents (Fig. 2.6).

This sensitive, selective and label-free nanopore method can perform simpler detection through electrical current measurement through nanopore than the optical and mass spectroscopy method. Therefore, it is suitable for real-time biosensing as a a ready-to-use device. As a candidate of portable biosensor, the aptamer-encoded nanopore has a high signal-to-noise ratio. The discrete signal change assures the detection of single target molecule at any time in case the sensing site is out of the laboratory or under electrical disturbance. Also, the characteristic of very strong binding between aptamer-encoded nanopore and target molecules allow to distinguish non-specific mal-detection by contaminants at the analytes from the interested detection; the contaminant would not produce stair-like current blocking but would rather produce spike-like current transient by passing through the nanopore without binding.

The new type of synthetic nanopore has more potential applications besides detecting bioterrorist agents. It can identify molecules of the pollutant or contaminator for water quality control. Also, it can be used for the purpose of medical diagnosis in quantifying biomarkers in blood samples. Furthermore, the aptamer-encoded glass-terminal nanopore could have large broad applications, since the aptamer is the key sensing element in the system and the new type of aptamer is still being developed *in vitro*. Therefore, the aptamer-encoded nanopore could evolve in any direction of usage as a new type of aptamer is developed. Another advantage of aptamer-based nanopore system is stability and durability of aptamers. They are superior in tolerance against most denaturing and degrading conditions than other biosensing elements such as proteins. Also, they are stable when immobilized to a specific site in the nanopore. In addition, aptamers could be synthesized, modified, and immobilized in a simple way. Thus, the device utilizing aptamers could be built with relatively low-cost, and could have various sensing characteristics through fine-tuning the aptamer for affinity and specificity. It is expected to devise the aptamer-encoded nanopore for detecting anthrax, another well-known bioterrorist agent, with aptamers developed to anthrax.

CONCLUSION

Nanopore research has developed very rapidly over a decade. The protein pore was first launched as a nanopore (Walker et al. 1992, Song et al. 1996) to bring the idea of DNA-sequencing through the nanopore (Akeson et al. 1999). Over a decade, research on nanopore has evolved dramatically, and has demonstrated various nanopore-related applications on, especially, biotechnology and pharmaceuticals. We have developed effective methods of investigating single-molecule folding/unfolding pathway trapped in the nanocavity of the pore. As micro- and nano-scale fabrication technique improved, it became possible to produce nanopore through the nano-fabrication technique. The synthetic nanopore joined later, but it has also evolved impressively fast with the superb properties on stability and controllability on pore size. Meanwhile, the biological nanopore was developed for a more stable system with support on the membrane. Within several years, the developing speed of nanopore has been boosted up for DNA-sequencing related applications, as it was envisioned since the biological nanopore presented its ability on detecting nano-scale molecules, by which the National Human Genome Research Institute (NHGRI) fosters US$1000 for the genome sequencing program. Nanopores have been developed in a short period of time with fruitful results, and are still to be raised up to the next level of applications as a research tool and biosensor. Again, the major advantage of nanopore is that it investigates and detects a single-molecule in a label-free environment. It is anticipated that the nanopore would contribute to the understanding of interactions and behavior of single molecules, and it would develop various bio-related sensors.

REFERENCES

Akeson, M. and D. Branton, J.J. Kasianowicz, E. Brandin and D.W. Deamer. 1999. Microsecond Time-Scale Discrimination Among Polycytidylic Acid, Polyadenylic Acid, and Polyuridylic Acid as Homopolymers or as Segments Within Single RNA Molecules. Biophys J 77: 3227–3233.

Akeson, M. and B. Hornblower, A. Coombs, R.D. Whitaker, A. Kolomeisky, S.J. Picone and A. Meller. 2007. Single-molecule analysis of DNA-protein complexes using nanopores. Nature Methods 4: 315–317.

Alberti, P. and J.L. Mergny. 2003. DNA duplex-quadruplex exchange as the basis for a nanomolecular machine. Proc Natl Acad Sci USA 100: 1569–1573.

Arthanari, H. and P.H. Bolton. 2001. Functional and dysfunctional roles of quadruplex DNA in cells. Chem Biol 8: 221–230.

Ashkenasy, N. and J. Sanchez-Quesada, H. Bayley and R. Ghadiri. 2005. Recognizing a Single Base in an Individual DNA Strand: A Step Toward DNA Sequencing in Nanopores. Angew Chem Int Ed 44:

Astier, Y. and O. Braha and H. Bayley. 2006. Toward single molecule DNA sequencing: Direct identification of ribonucleoside and deoxyribonucleoside 5 '-monophosphates by using an engineered protein nanopore equipped with a molecular adapter. J Am Chem Soc 128: 1705–1710.

Baskerville, S. and M. Zapp and A.D. Ellington. 1999. Anti-Rex aptamers as mimics of the Rex-binding element. J Virol 73: 4962–4971.

Bayley, H. and P.S. Cremer. 2001. Stochastic sensors inspired by biology. Nature 413: 226–230.

Bayley, H. and L. Jayasinghe. 2004. Functional engineered channels and pores (Review). Mol Membr Biol 21: 209–220.

Bock, L.C. and L.C. Griffin, J.A. Latham, E.H. Vermaas and J.J. Toole. 1992. Selection of Single-Stranded-DNA Molecules That Bind and Inhibit Human Thrombin. Nature 355: 564–566.

Braha, O. and L.Q. Gu, L. Zhou, X.F. Lu, S. Cheley and H. Bayley. 2000. Simultaneous stochastic sensing of divalent metal ions. Nature Biotechnol 18: 1005–1007.

Branton, D. and D.W. Deamer, A. Marziali, H. Bayley, S.A. Benner, T. Butler, M. Di Ventra, S. Garaj, A. Hibbs, X.H. Huang, S.B. Jovanovich, P.S. Krstic, S. Lindsay, X.S.S. Ling, C.H. Mastrangelo, A. Meller, J.S. Oliver, Y.V. Pershin, J.M. Ramsey, R. Riehn, G.V. Soni, V. Tabard-Cossa, M. Wanunu, M. Wiggin and J.A. Schloss. 2008. The potential and challenges of nanopore sequencing. Nature Biotechnol 26: 1146–1153.

Burge, S. and G.N. Parkinson, P. Hazel, A.K. Todd and S. Neidle. 2006. Quadruplex DNA: sequence, topology and structure. Nucleic Acids Res 34: 5402–5415.

Butler, T.Z. and M. Pavlenok, I.M. Derrington, M. Niederweis and J.H. Gundlach. 2008. Single-molecule DNA detection with an engineered MspA protein nanopore. Poc Natl Acad Sci USA 105: 20647–20652.

Capone, R. and S. Blake, M.R. Restrepo, J. Yang and M. Mayer. 2007. Designing nanosensors based on charged derivatives of gramicidin A. J Am Chem Soc 129: 9737–9745.

Cheley, S. and L-Q. Gu and H. Bayley. 2002. Stochastic Sensing of Nanomolar Inositol 1,4,5-Trisphosphate with an Engineered Pore. Chem Biol 9: 829–838.

Clarke, J. and H.C. Wu, L. Jayasinghe, A. Patel, S. Reid and H. Bayley. 2009. Continuous base identification for single-molecule nanopore DNA sequencing. Nature Nanotechnology 4: 265–270.

Cornell, B.A. and V.L.B. BraachMaksvytis, L.G. King, P.D.J. Osman, B. Raguse, L. Wieczorek and R.J. Pace. 1997. A biosensor that uses ion-channel switches. Nature 387: 580–583.

Costello, R.F. and I.P. Peterson, J. Heptinstall, N.G. Byrne and L.S. Miller. 1998. A robust gel-bilayer channel biosensor. Adv Mater Opt Electr 8: 47–52.

Davis, J.T. and G.P. Spada. 2007. Supramolecular architectures generated by self-assembly of guanosine derivatives. Chem Soc Rev 36: 296–313.

de Soultrait, V.R. and P.Y. Lozach, R. Altmeyer, L. Tarrago-Litvak, S. Litvak and M.L. Andreola. 2002. DNA aptamers derived from HIV-1 RNase H inhibitors are strong anti-integrase agents. J Mol Biol 324: 195–203.

Deamer, D.W. and D. Branton. 2002. Characterization of Nucleic Acids by Nanopore Analysis. Acc Chem Res 35: 817–825.

Derrington, I.M. and T.Z. Butler, M.D. Collins, E. Manrao, M. Pavlenok, M. Niederweis and J. H. Gundlach. 2010. Nanopore DNA sequencing with MspA. Poc Natl Acad Sci USA 107: 16060–16065.

Ding, S. and C.L. Gao and L.Q. Gu. 2009. Capturing Single Molecules of Immunoglobulin and Ricin with an Aptamer-Encoded Glass Nanopore. Anal Chem 81: 6649–6655.

Ellington, A.D. and J.W. Szostak. 1990. *In vitro* Selection of Rna Molecules That Bind Specific Ligands. Nature 346: 818–822.

Fertig, N. and C. Meyer, R.H. Blick, C. Trautmann and J.C. Behrends. 2001. Microstructured glass chip for ion-channel electrophysiology. Phys Rev E 64: 040901.

Fertig, N. and R.H. Blick and J.C. Behrends. 2002. Whole Cell Patch Clamp Recording Performed on a Planar Glass Chip. Biophys J 82: 3056–3062.

Gao, C.L. and S. Ding, Q.L. Tan and L. Q. Gu. 2009. Method of Creating a Nanopore-Terminated Probe for Single-Molecule Enantiomer Discrimination. Anal Chem 81: 80–86.

Garaj, S. and W. Hubbard, A. Reina, J. Kong, D. Branton and J.A. Golovchenko. 2010. Graphene as a subnanometre trans-electrode membrane. Nature 467: 190–U173.

Gu, L. Q. and O. Braha, S. Conlan, S. Cheley and H. Bayley. 1999. Stochastic sensing of organic analytes by a pore-forming protein containing a molecular adapter. Nature 398: 686–690.

Han, A.P. and G. Schurmann, G. Mondin, R.A. Bitterli, N.G. Hegelbach, N.F. de Rooij and U. Staufer. 2006. Sensing protein molecules using nanofabricated Pores. Appl Phys Lett 88: 093901.

Hardin, C.C. and A.G. Perry and K. White. 2000. Thermodynamic and kinetic characterization of the dissociation and assembly of quadruplex nucleic acids. Biopolymers 56: 147–194.

Heng, J.B. and C. Ho, T. Kim, R. Timp, A. Aksimentiev, Y.V. Grinkova, S. Sligar, K. Schulten and G. Timp. 2004. Sizing DNA using a nanometer-diameter pore. Biophys J 87: 2905–2911.

Heng, J.B. and A. Aksimentiev, C. Ho, P. Marks, Y.V. Grinkova, S. Sligar, K. Schulten and G. Timp. 2005. Stretching DNA using the electric field in a synthetic nanopore. Nano Lett 5: 1883–1888.

Heyduk, T. and E. Heyduk. 2002. Molecular beacons for detecting DNA binding proteins. Nature Biotechnol 20: 171–176.

Hille, B. 2001. Ion Channels of Excitable Membranes.

Holden, M.A. and L. Jayasinghe, O. Daltrop, A. Mason and H. Bayley. 2006. Direct transfer of membrane proteins from bacteria to planar bilayers for rapid screening by single-channel recording. Nature Chem Biol 2: 314–318.

Howorka, S. and Z. Siwy. 2009. Nanopore analytics: sensing of single molecules. Chem Soc Rev 38: 2360–2384.

Jeon, S.H. and B. Kayhan, T. Ben-Yedidia and R. Arnon. 2004. A DNA aptamer prevents influenza infection by blocking the receptor binding region of the viral hemagglutinin. J Biol Chem 279: 48410–48419.

Jeon, T.J. and N. Malmstadt and J.J. Schmidt. 2006. Hydrogel-encapsulated lipid membranes. J Am Chem Soc 128: 42–43.

Jing, N.J. and R.F. Rando, Y. Pommier and M.E. Hogan. 1997. Ion selective folding of loop domains in a potent anti-HIV oligonucleotide. Biochemistry 36: 12498–12505.

Kang, X.F. and S. Cheley, X.Y. Guan and H. Bayley. 2006. Stochastic detection of enantiomers. J Am Chem Soc 128: 10684–10685.

Kang, X.F. and S. Cheley, A.C. Rice-Ficht and H. Bayley. 2007. A storable encapsulated bilayer chip containing a single protein nanopore. J Am Chem Soc 129: 4701–4705.

Kasianowicz, J.J. and E. Brandin, D. Branton and D.W. Deamer. 1996. Characterization of individual polynucleotide molecules using a membrane channel. Poc Natl Acad Sci U S A 93: 13770–13773.

Klajn, R. and M. Fialkowski, I.T. Bensemann, A. Bitner, C.J. Campbell, K. Bishop, S. Smoukov and B.A. Grzybowski. 2004. Multicolour micropatterning of thin films of dry gels. Nature Mater 3: 729–735.

Klug, S.J. and A. Huttenhofer and M. Famulok. 1999. *In vitro* selection of RNA aptamers that bind special elongation factor SelB, a protein with multiple RNA-binding sites, reveals one major interaction domain at the carboxyl terminus. RNA 5: 1180–1190.

Krapf, D. and M.Y. Wu, R.M.M. Smeets, H.W. Zandbergen, C. Dekker and S.G. Lemay. 2006. Fabrication and characterization of nanopore-based electrodes with radii down to 2 nm. Nano Lett 6: 105–109.

Li, J. and D. Stein, C. McMullan, D. Branton, M.J. Aziz and J.A. Golovchenko. 2001. Ion-beam sculpting at nanometre length scales. Nature 412: 166–169.

Li, J.W.J. and W.H. Tan. 2002. A single DNA molecule nanomotor. Nano Lett 2: 315–318.

Li, J.L. and M. Gershow, D. Stein, E. Brandin and J.A. Golovchenko. 2003. DNA molecules and configurations in a solid-state nanopore microscope. Nature Mater 2: 611–615.

Luchian, T. and S.H. Shin and H. Bayley. 2003. Single-Molecule Covalent Chemistry with Spatially Separated Reactants. Angew Chem Int Ed 42: 3766–3771.

Malmstadt, N. and M.A. Nash, R.F. Purnell and J.J. Schmidt. 2006. Automated formation of lipid-bilayer membranes in a microfluidic device. Nano Lett 6: 1961–1965.

Marathias, V.M. and P.H. Bolton. 2000. Structures of the potassium-saturated, 2 : 1, and intermediate, 1 : 1,forms of a quadruplex DNA. Nucleic Acids Res 28: 1969–1977.

Martin, C.R. and Z.S. Siwy. 2007. Learning nature's way: Biosensing with synthetic nanopores. Science 317: 331–332.

Mathe, J. and H. Visram, V. Viasnoff, Y. Rabin and A. Meller. 2004. Nanopore unzipping of individual DNA hairpin molecules. Biophys J 87: 3205–3212.

Mayer, M. and J. Yang, I. Gitlin, D.H. Gracias and G.M. Whitesides. 2004. Micropatterned agarose gels for stamping arrays of proteins and gradients of proteins. Proteomics 4: 2366–2376.

McNally, B. and M. Wanunu and A. Meller. 2008. Electromechanical Unzipping of Individual DNA Molecules Using Synthetic Sub-2 nm Pores. Nano Lett 8: 3418–3422.

Meller, A. and L. Nivon, E. Brandin, J. Golovchenko and D. Branton. 2000. Rapid nanopore discrimination between single polynucleotide molecules. Proc Natl Acad Soc USA 97: 1079–1084.

Meller, A. and L. Nivon and D. Branton. 2001. Voltage-driven DNA translocations through a nanopore. Phys Rev Lett 86: 3435–3438.

Merchant, C.A. and K. Healy, M. Wanunu, V. Ray, N. Peterman, J. Bartel, M.D. Fischbein, K. Venta, Z.T. Luo, A.T.C. Johnson and M. Drndic. 2010. DNA Translocation through Graphene Nanopores. Nano Lett 10: 2915–2921.

Mitchell, N. and S. Howorka. 2008. Chemical tags facilitate the sensing of individual DNA strands with nanopores. Angew Chem Int Ed 47: 5565–5568.

Montal, M. and P. Mueller. 1972. Formation of Biomolecular Membranes from Lipid Monolayers and a study of Their Electrical Properties. Proc Natl Acad Soc USA 69: 3561–3566.

Movileanu, L. and S. Howorka, O. Braha and H. Bayley. 2000. Detecting protein analytes that modulate transmembrane movement of a polymer chain within a single protein pore. Nature Biotechnol 18: 1091–1095.

Nakane, J. and M. Wiggin and A. Marziali. 2004. A nanosensor for transmembrane capture and identification of single nucleic acid molecules. Biophys J 87: 615–621.

Naumann, C.A. and O. Prucker, T. Lehmann, J. Ruhe, W. Knoll and C.W. Frank. 2002. The polymer-supported phospholipid bilayer: Tethering as a new approach to substrate-membrane stabilization. Biomacromolecules 3: 27–35.

Neidle, S. and G. Parkinson. 2002. Telomere maintenance as a target for anticancer drug discovery. Nature Rev Drug Discov 1: 383–393.

Padmanabhan, K. and K.P. Padmanabhan, J.D. Ferrara, J.E. Sadler and A. Tulinsky. 1993. The Structure of Alpha-Thrombin Inhibited by a 15-Mer Single-Stranded-DNA Aptamer. J Biol Chem 268: 17651–17654.

Pluen, A. and P.A. Netti, R.K. Jain and D.A. Berk. 1999. Diffusion of macromolecules in agarose gels: Comparison of linear and globular configurations. Biophys J 77: 542–552.

Sackmann, E. 1996. Supported Membranes: Scientific and Practical Applications. Science 271: 43–48.

Saleh, O.A. and L.L. Sohn. 2003. Direct detection of antibody-antigen binding using an on-chip artificial pore. Proc Natl Acad Soc USA 100: 820–824.

Schmidt, C. and M. Mayer and H. Vogel. 2000. A chip-based biosensor for the functional analysis of single ion channels. Angew Chem Int Ed 39: 3137–3140.

Schneider, G.F. and S. W. Kowalczyk, V.E. Calado, G. Pandraud, H.W. Zandbergen, L.M.K. Vandersypen and C. Dekker. 2010. DNA Translocation through Graphene Nanopores. Nano Lett 10: 3163–3167.

Sen, D. and W. Gilbert. 1988. Formation of Parallel 4-Stranded Complexes by Guanine-Rich Motifs in DNA and Its Implications for Meiosis. Nature 334: 364–366.

Sexton, L.T. and L.P. Horne, S.A. Sherrill, G.W. Bishop, L.A. Baker and C.R. Martin. 2007. Resistive-pulse studies of proteins and protein/antibody complexes using a conical nanotube sensor. J Am Chem Soc 129: 13144–13152.

Shim, J.W. and L.Q. Gu. 2007. Stochastic sensing on a modular chip containing a single-ion channel. Anal Chem 79: 2207–2213.

Shim, J.W. and L.Q. Gu. 2008. Encapsulating a single G-quadruplex aptamer in a protein nanocavity. J Phys Chem B 112: 8354–8360.

Shim, J.W. and Q.L. Tan and L.Q. Gu. 2009. Single-molecule detection of folding and unfolding of the G-quadruplex aptamer in a nanopore nanocavity. Nucleic Acids Res 37: 972–982.

Siwy, Z. and A. Fulinski. 2002. Fabrication of a synthetic nanopore ion pump. Phys Rev Lett 89: 198103.

Siwy, Z. and L. Trofin, P. Kohli, L.A. Baker, C. Trautmann and C.R. Martin. 2005. Protein biosensors based on biofunctionalized conical gold nanotubes. J Am Chem Soc 127: 5000–5001.

Song, L. and M.R. Hobaugh, C. Shustak, S. Cheley, H. Bayley and J.E. Gouaux. 1996. Structure of Staphylococcal alpha -Hemolysin, a Heptameric Transmembrane Pore. Science 274: 1859–1865.

Stoddart, D. and A.J. Heron, E. Mikhailova, G. Maglia and H. Bayley. 2009. Single-nucleotide discrimination in immobilized DNA oligonucleotides with a biological nanopore. Proc Natl Acad Soc USA 106: 7702–7707.

Storm, A.J. and J.H. Chen, X.S. Ling, H.W. Zandbergen and C. Dekker. 2003. Fabrication of solid-state nanopores with single-nanometre precision. Nature Mater 2: 537–540.

Storm, A.J. and J.H. Chen, H.W. Zandbergen and C. Dekker. 2005. Translocation of double-strand DNA through a silicon oxide nanopore. Phys Rev E 71: 051903.

Tang, Z.W. and D. Shangguan, K.M. Wang, H. Shi, K. Sefah, P. Mallikratchy, H.W. Chen, Y. Li and W.H. Tan. 2007. Selection of aptamers for molecular recognition and characterization of cancer cells. Anal Chem 79: 4900–4907.

Tropini, C. and A. Marziali. 2007. Multi-nanopore force Spectroscopy for DNA analysis. Biophys J 92: 1632–1637.

Tuerk, C. and L. Gold. 1990. Systematic Evolution of Ligands by Exponential Enrichment—Rna Ligands to Bacteriophage-T4 DNA-Polymerase. Science 249: 505–510.

Uram, J.D. and K. Ke, A.J. Hunt and M. Mayer. 2006. Label-free affinity assays by rapid detection of immune complexes in submicrometer pores. Angew Chem Int Ed 45: 2281–2285.

Venkatesan, B.M. and A.B. Shah, J.M. Zuo and R. Bashir. 2010. DNA Sensing Using Nanocrystalline Surface-Enhanced Al2O3 Nanopore Sensors. Adv Funct Mater 20: 1266–1275.

Vercoutere, W. and S. Winters-Hilt, H. Olsen, D. Deamer, D. Haussler and M. Akeson. 2001. Rapid discrimination among individual DNA hairpin molecules at single-nucleotide resolution using an ion channel. Nature Biotechnol 19: 248–252.

Walker, B. and M. Krishnasastry, L. Zorn and H. Bayley. 1992. Assembly of the Oligomeric Membrane Pore Formed by Staphylococcal Alpha-Hemolysin Examined by Truncation Mutagenesis. J Biol Chem 267: 21782–21786.

Wen, J.D. and D.M. Gray. 2004. Selection of genomic sequences that bind tightly to Ff gene 5 protein: primer-free genomic SELEX. Nucleic Acids Res 32: e182.

Xiao, Y. and A.A. Lubin, A.J. Heeger and K.W. Plaxco. 2005. Label-free electronic detection of thrombin in blood serum by using an aptamer-based sensor. Angew Chem Int Ed 44: 5456–5459.

Zandbergen, H.W. and R.J.H.A. van Duuren, P.F.A. Alkemade, G. Lientschnig, O. Vasquez, C. Dekker and F.D. Tichelaar. 2005. Sculpting nanoelectrodes with a transmission electron beam for electrical and geometrical characterization of nanoparticles. Nano Lett 5: 549–553.

3

Self-assembled Peptide and Protein Nanostructures in Diagnosis

Jaime Castillo-León,[1,a,*] *Yuri Belotti*[1] and *Winnie E. Svendsen*[1]

INTRODUCTION

In the medical field to be able to diagnose disease as early as possible in a reproducible, rapid, sensitive, selective, low-cost and non-painful way is close to a dream. An early detection of such disease markers as possible will allow a faster reaction time for finding the right treatment, thereby stopping the advance of a specific disease, e.g., cancer.

Researchers in different fields are constantly making efforts to develop diagnostic devices capable of fulfilling—if not all, then as many as possible—of the above-mentioned characteristics. In this respect, nanotechnology plays a major role on the long path toward the manufacture of efficient diagnostic platforms. Advances in micro- and nano-fabrication, synthesis and characterization of new materials, integration of several techniques in a single device, decreases in sample and reagent amounts are some of the advantages that make nanotechnology such a powerful tool in the biomedical field.

New materials in the nanometer range are constantly being synthesized and characterized in order to evaluate their possible use in various biomedical applications such as drug delivery, tissue engineering, biosensor development or the maturation of diagnostic platforms, among

[1]Department of Micro and Nanotechnology, Technical University of Denmark Building 345 east, Lyngby 2800, Denmark.

[a]E-mail: jaic@nanotech.dtu.dk

many others. Traditionally nanomaterials such as carbon nanotubes, silicon nanowires or metallic nanoparticles have been widely used due to their advantages in terms of electrical and optical properties as well as their easy functionalization and integration in biosensing platforms (Ahmed 2008, He 2008, Perez-Lopez and Merkoci 2011, Qureshi 2009, Ramgir 2010, Siangproh 2011, Wang 2005). However, in some cases, a few of these nanomaterials are difficult and expensive to fabricate or present disadvantages in terms of biocompatibility. In order to overcome these challenges, researchers are constantly exploring and testing new materials obtained from a large variety of sources: proteins, peptides, oxides, etc. (Kim 2010).

Recently, self-assembled peptides and proteins have emerged as promising materials for applications in biomedical fields (de la Rica and Matsui 2010, Gao and Matsui 2005, Hauser and Zhang 2010, Toksoz 2010). These biomaterials can be employed to fabricate 1D, 2D or 3D nanostructures in a very facile way, without involving complicated machinery or using extreme temperatures (Reches and Gazit 2004, Reches and Gazit 2006a). Structures such as nanotubes, nanowires, nanoparticles or hydrogels have been synthesized utilizing self-assembled peptides and proteins as building blocks.

This chapter presents and discusses the use of such biological nanomaterials for the synthesis of nanostructures in the development of diagnostic devices. The advantages and challenges involving the utilization of these materials as well as their applications in diagnostics are highlighted.

SELF-ASSEMBLED PEPTIDES AND PROTEINS

In nature, building blocks such as peptides and proteins are able to self-organize in efficient and ordered assembly processes. These bottom-up processes take place rapidly under mild conditions, such as in aqueous environments, at room temperature. Biological materials such as elastin, silk, collagen, short aromatic peptides or polypeptides are capable of self-assembly thus creating nanostructures with very attractive features for use in bionanotechnology applications. Compared to traditional nanomaterials (e.g., carbon nanotubes, silicon nanowires, metallic nanoparticles, etc.), they offer several advantages in terms of fabrication (i.e., low cost, short time and simplicity), functionalization and biocompatibility.

The self-organization processes resulting in the creation of nanostructures of varying shapes undergo various mechanisms such as hydrophobic or hydrophilic interactions, charge attraction or repulsion, or stacking of building blocks into beta sheets. An example of the several types of nanostructures obtained by the self-assembly of peptides can be found in Fig. 3.1, where the various nanostructures formed by a short aromatic dipeptide, diphenylalanine, are shown.

A detailed discussion of these processes is beyond the scope of this chapter; however Table 3.1 presents examples of several building blocks used to fabricate supramolecular structures by self-assembly as well as the driving force behind the formation of the structures.

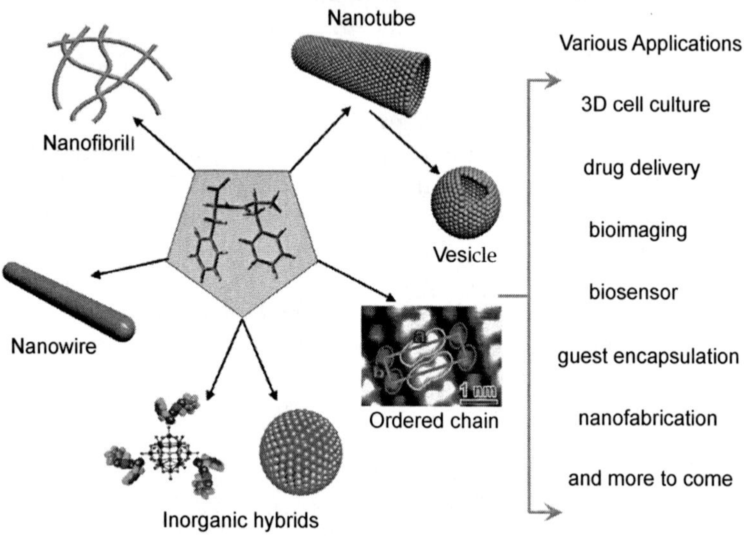

Figure 3.1: Representation of the various structures formed by self-assembly of diphenylalanine building blocks and their potential applications in nanotechnology. (Yan et al. 2010. Self-assembly and application of diphenylalanine-based nanostructures. Chem Soc Rev 2010, 39, 1877–1890. Reproduced by permission of The Royal Society of Chemistry).

Color image of this figure appears in the color plate section at the end of the book.

ADVANTAGES AND CHALLENGES OF EMPLOYING SELF-ASSEMBLED PEPTIDES AND PROTEINS

As previously mentioned, the fabrication of nanostructures by the self-assembly of biological building blocks is a rapid process that takes place under non-aggressive conditions. On the other hand, the fabrication of carbon nanotubes, silicon nanowires or metallic nanoparticles used in the development of biosensing devices requires specialized equipment and high temperatures. In some cases, the fabrication takes place in a clean-room. These requirements increase the cost and fabrication time of the desired structures. The self-assembly of nanometric configurations

Table 3.1: Biological building blocks for the synthesis of self-assembled nanostructures and the driving force for the formation of the structures.

Building block	**Driving force**	**Nanostructure**	**Reference**
EAK16-II	Hydrophobic-hydrophilic interactions	Nanofibres	(Zhang 1993)
Diphenylalanine	Stacking interaction between aromatic moieties	Nanotubes and nanoparticles	(Reches and Gazit 2004)
Cyclo[-(D-Ala-Glu-D-Ala-Gln)2]	Hydrogen bonds interactions	Nanotubes	(Ghadiri 1993)
Bolaamphiphiles peptide-conjugate	Intermolecular association of hydrophobic moieties in aqueous solution	Nanotubes	(Matsui 2001)
Arginine-Alanine-Aspartate RAD16-I and RAD16-II	Ionic complementary self-assembly	Nanofibres	(Holmes 2000)
Diacetylene macromonomers	Solvophobic interactions combined with hydrogen bonding	Helical nanotapes, ribbons and double ribbons	(Frauenrath and Jahnke 2008)
N-fluorenylmethoxycarbonyl diphenylalanine	π–π stacking and hydrogen bonding	Hydrogel	(Smith et al. 2008)
Peptide TZ1H	Enzyme-assisted self-assembly	Helical fibrils	(Zimenkov et al. 2006)
11-mer ring protein trp RNA-binding attenuation protein	Dissulphide bonds and hydrophobic interactions	Nanotube	(Miranda et al. 2009)

using self-assembled peptides or proteins offers an alternative to carbon nanotubes, silicon nanowires or metallic particles in the manufacture of devices for diagnosis.

One example involves the synthesis of biological nanotubes based on a short aromatic peptide, diphenylalanine. This process takes only a couple of seconds and is carried out in an aqueous environment at room temperature (Yan 2010). Moreover, the on-chip fabrication of self-assembled peptide nanotubes and nanoparticles was recently demonstrated by our group. In less than a minute, hundreds of structures were obtained in a very fast and simple manner (Castillo-León et al. 2011). Daamen and co-workers presented the preparation of elastin nanocapsules with a high potential for use in various applications in nanomedicine, e.g., drug delivery, imaging or vaccination (Daamen et al. 2007). Apart from their rapid, simple and low-cost fabrication, self–assembled peptides and proteins present great advantages in terms of functionalization. The nanostructures synthesized with biological building blocks can be readily decorated with functional compounds such as enzymes, antibodies, fluorescent molecules or metallic nanoparticles.

Reches and Gazit showed how nanotubes could be formed by the self-assembly of a short aromatic dipeptide, i.e., the diphenylalanine peptide, with biotin moieties in order to enable the selective functionalization of the nanometric structures with avidin-labelled species (Reches and Gazit 2007). Based on this method, peptide nanotubes were functionalized with gold nanoparticles, quantum dots and fluorescent molecules. Figure 3.2 illustrates peptide nanotubes decorated with a fluorescent molecule and with quantum dots following the method of Reches and Gazit.

The integration of these functional groups with the biological nanostructures opens for a large spectrum of possibilities involving the use of the functionalized supramolecular structures in the development of biosensing devices, drug delivery systems and imaging agents, among others.

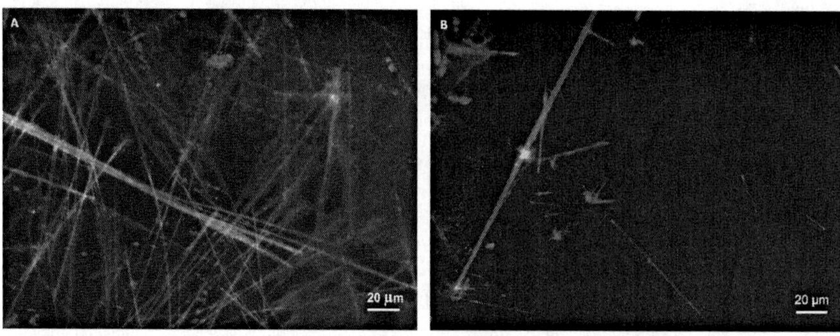

Figure 3.2: Labelling of self-assembled peptide nanotubes with (A) a fluorescent molecule, Atto-610, and (B) quantum dots via avidin-streptavidin interaction.

Although biological self-assembled nanostructures have been extensively used in medicine in a large number of applications (de la Rica and Matsui 2010, Yan 2010, Gras and Rudy 2009), this chapter focuses exclusively on the use of these nanostructures in diagnostic applications. Despite all the advantages of working with this biological material in the development of diagnostic devices, several challenges need to be confronted and solved. These challenges are related to controlling the size during the synthesis and manipulating the nanostructures in order to integrate them with transducers. Regulating the dimension of biological self-assembled nanostructures is not an easy task due to the spontaneous nature of the self-assembly process. Nevertheless, by controlling the size of the self-assembled structures, one can develop sensing devices of similar dimensions resulting in more reproducible devices.

Several methods have been used in order to control the final size of synthesized self-assembled nanostructures. Usui and co-workers proposed a novel strategy for producing size-regulated filament-shaped structures.

This strategy is called Nanolego and consists of utilizing building elements of a structurally stable symmetrical homo-oligomeric protein and binding elements constituted of multiple heterointeraction proteins with a relatively weak affinity (Usui et al. 2009). In a different approach, Porrata and co-workers synthesized self-assembled peptide nanotubes using polycarbonate membranes in solution as templates to control the diameter of the nanotubes (Porrata et al. 2002). With this method, peptide nanotubes as small as 50 nm in diameter were self-assembled in the membrane pores, as depicted in Fig. 3.3.

The second important challenge when working with nanometric structures of self-assembled proteins or peptides has to do with their manipulation, which in addition to integration is necessary in order to connect them with transducers for the development of sensing platforms used in diagnostics. However, the linking of the macroscopic world to the nanoscopic world of biological nanostructures is a technological challenge.

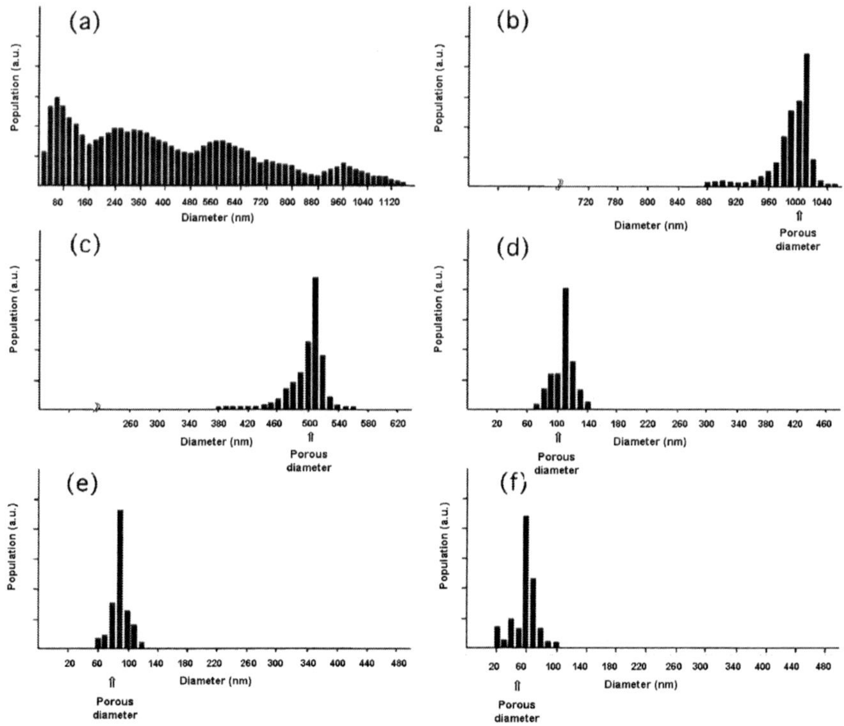

Figure 3.3: Diameter distribution of templated self-assembled peptide nanotubes synthesized in (a) suspension; (b) 1-μm membrane; (c) 500-nm membrane; (d) 100-nm membrane; (e) 80-nm membrane; and (f) 50-nm membrane. The arrows mark the polycarbonate membrane pore sizes. Reprinted with permission from Porrata et al. 2002. Chem Mater 14, 4378–4381. Copyright 2002 American Chemical Society.

Various methods have been developed in order to overcome this challenge in such a way that the manipulation can be done without affecting the structure or without altering the properties of the biomaterial.

Zhao and Matsui developed an accurate immobilization method via biological molecular recognition to manipulate and place antibody-functionalized peptide nanotubes on top of protein-patterned arrays. Based on this method, which is depicted in Fig. 3.4, Zhao and Matsui were able to increase the yield of nanotube attachment by maximizing the antibody-antigen interaction (Zhao and Matsui 2007, Zhao et al. 2005).

Dielectrophoresis is a technique where a polarized particle suspended in a non-homogeneous electric field can be oriented and connected to electrodes due to the induced forces on each half of the dipole. This technique was used for the manipulation and immobilization of self-

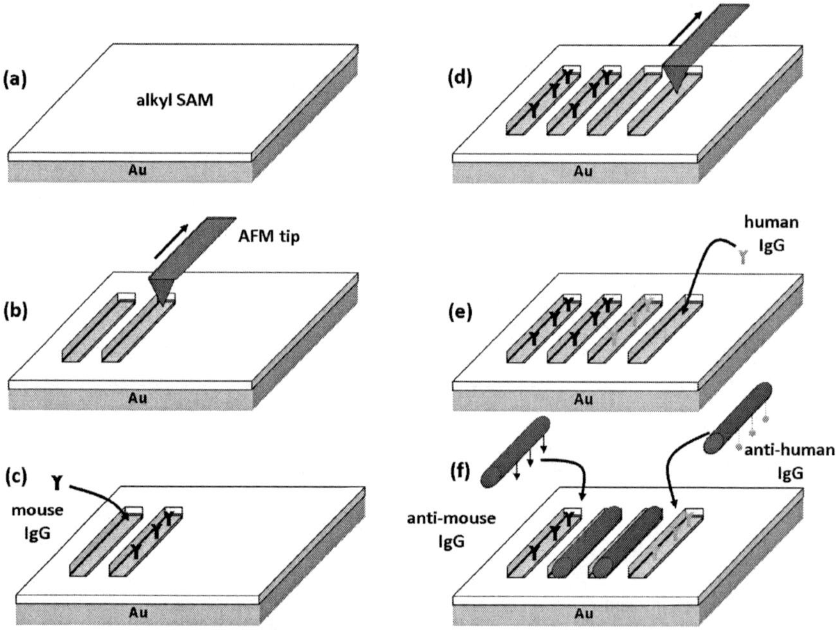

Figure 3.4: A schematic diagram to assemble anti-mouse IgG-coated nanotubes and anti-human IgG-coated nanotubes onto their antigen-patterned substrates via biological recognition. (a) Self-assembly of alkylthiol monolayers on Au substrates. (b) Shaving trenches on the alkylthiol SAM with an AFM tip. (c) Deposition of mouse IgG in the shaved trenches. (d) Shaving another array of trenches on the alkylthiol SAM using an AFM tip. (e) Deposition of human IgG in the shaved trenches. (f) Location-specific immobilization of Alexa Fluor 546-labeled anti-mouse IgG nanotubes into the mouse IgG trenches and FITC-labelled anti-human IgG nanotubes into the human IgG trenches via their biological recognition. Reprinted with permission from Zhao, Z. et al. 2005. J Am Chem Soc 127, 8930–8931. Copyright 2005 American Chemical Society.

Color image of this figure appears in the color plate section at the end of the book.

assembled peptide nanotubes on top of gold electrodes (Castillo et al. 2008, de la Rica et al. 2008). In this way, it was possible to characterize their electrical properties and to fabricate a label-free pathogen sensor assembled from peptide nanotubes (de la Rica et al. 2008).

A third example of a controlled manipulation of self-assembled biological nanostructures has been presented by Reches and Gazit. Their work involved achieving a horizontal alignment of biological nanotubes with a ferrofluid and applying an external magnetic field. Their results demonstrate the ability to form a two-dimensional dense array of peptide nanotube assemblies with either vertical or horizontal patterns (Reches and Gazit 2006b). Another option for the manipulation of self-assembled protein or peptide nanostructures is the use of laser tweezers. In a study by Kumara et al., flagella protein nanotubes with cysteine loops with diameters below 50 nm were optically trapped using a biologically compatible infrared wavelength (1064 nm) laser tweezer (Kumara et al. 2006).

These four examples are only a few of the many developed in order to carry out manipulation in a controlled manner and without altering the natural properties of these biological entities. Many other techniques such as atomic force microscopy or microfluidics have been employed for manipulation purposes, however a more detailed description of these and other available methods are beyond the scope of this chapter. The reader is referred to available articles and books dealing with this topic (Castillo et al. 2009, Castillo et al. 2011, Sedman et al. 2009).

DIAGNOSTIC DEVICES USING SELF-ASSEMBLED PROTEIN AND PEPTIDE NANOSTRUCTURES

This section focuses on the use of protein and peptide-based nanostructures such as tubes, fibres, particles, or gels for the development of diagnostic devices with applications in medicine. These types of well-defined biological structures are integrated with transducers that will be used for the detection of compounds of biomedical relevance based on changes in current, potential, or light intensity among other factors. The section is divided according to the type of parameter used to assess the presence or concentration of the biomedical compound interacting with the functionalized self-assembled structure. Some of the most relevant investigations involving the use of self-assembling proteins and peptides are reviewed, mentioning details of functionalization and application. The section ends with a table summarizing some of the available diagnostic devices in order to provide the reader with an idea of the different opportunities and the flexibility of these types of nanostructures.

Electrochemical and Electrical Diagnostic Devices

As previously mentioned, nanostructures of self-assembling peptides and proteins are not very conductive materials (Castillo et al. 2008, Scheibel et al. 2003). Effective functionalization strategies are thus necessary in order to increase the conductivity of the biological configurations, thereby allowing interaction with the sample and the resultant change in current or potential. Functional compounds such as enzymes, antibodies and conducting polymers have been employed to obtain interactions between the 3D self-assembled structure and the compound to be detected.

The modification of self-assembled peptide nanotubes with glucose oxidase (GOx) and ethanol dehydrogenase (HPOG) for the electrochemical detection of glucose, and ethanol has been presented by Yemini and co-workers. These modified nanostructures were immobilized using Traut's reagent on top of gold electrodes, and polyethylenimine with glutaraldehyde was used as a immobilization matrix to keep the enzymes and the tubes on top of the transducer (Yemini et al. 2005a). A similar approach was developed by Yang et al. It involved integrating glucose oxidase into an ionic-complementary peptide nanofibre matrix. The nanofibres were synthesized using the ionic-complementary peptide EAK16-II. The response of the developed sensor displayed a sensitivity of 26 nA mM^{-1} that was linear up to 10 mM glucose. This covered the range 3.5–6.5 mM required for clinical applications (Yang et al. 2009). Figure 3.5 displays the procedure of enzymatic electrode fabrication and its sensing mechanism.

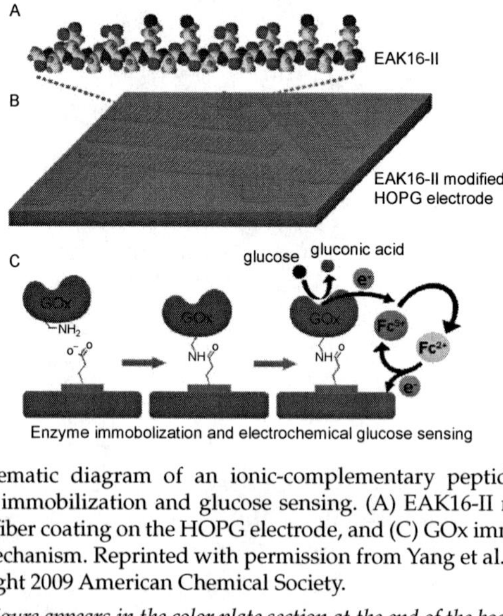

Figure 3.5: A schematic diagram of an ionic-complementary peptide modified HOPG electrode for GOx immobilization and glucose sensing. (A) EAK16-II molecular structure, (B) EAK16-II nanofiber coating on the HOPG electrode, and (C) GOx immobilization and the glucose sensing mechanism. Reprinted with permission from Yang et al. 2009. Langmuir, 25, 7773–7777. Copyright 2009 American Chemical Society.

Color image of this figure appears in the color plate section at the end of the book.

Diphenylalanine was used to synthesize nanotubes that were combined with micro-peroxidase-11 and immobilized onto Indium Tin Oxide (ITO) electrodes for the detection of hydrogen peroxide. Poly(allylamine hydrochloride) was selected as the immobilization matrix. In this way, the amperometric detection of hydrogen peroxide was performed obtaining a sensitivity of 9.43 $\mu A\ cm^{-2}\ mmol^{-1}\ L$ and a detection limit of 6 $\mu mol\ L^{-1}$ (Cipriano et al. 2010).

The detection of *Escherichia coli* O157:H7, a food borne pathogenic bacteria, was made possible by using self-assembled peptide nanotubes synthesized with cyclic peptide (cyclo[(Gln-D-Leu)$_4$]. The nanotubes were functionalized with antibodies against *Escherichia coli* O157:H7. The functionalized tubes were deposited onto screen-printed carbon paste electrodes (Cho et al. 2008).

Viguier et al. 2011 described the fabrication of an amperometric biosensor for the detection of copper in solution, which is associated with two severe disorders known as Wilson disease and Menkes disease (Huster and Lutsenko 2007). The work involved screen printed gold electrodes modified with self-assembled peptide nanofibres synthesized from an octapeptide with cysteine substitutes. In this system, the interaction between the nanofibre and the copper ions was due to the presence of donor atoms at the peptide backbone which were able to complexate to metal ions. A linear range up to 50 µM copper and a sensitivity of 0.68 $\mu A\ cm^{-2}\ \mu M^{-1}$ was obtained (Viguier et al. 2011).

The herpes simplex virus type 2 (HSV-2) was detected using a label-free sensor chip assembled from peptide nanotubes. These tubes were prepared by self-assembly from bolaamphiphilic peptide monomers and then modified with antibodies against HSV-2. The modified antibodies were subsequently immobilized to the gap between a pair of gold electrodes by dielectrophoresis. The binding of the nanoscale viral particles to the modified peptide nanotubes was detected on the basis of a capacitance change (de la Rica et al. 2008). This system was extended by the same group to the multiplexed detection of bacteria with a reusable pathogen biochip (de la Rica et al. 2010). Using this sensor chip, *E. coli* and *S. typhi* cells were detected in the range from 10^2 to 10^4 cells within an hour. In this approach, the biological nanotubes were functionalized with anti-*E. coli* and IgG antibodies. Figure 3.6 displays a scheme of the sensor chip, a SEM image of the nanotubes and impedance measurements.

Finally, an interesting system was developed by Sasso et al., where self-assembled peptide nanowires were modified with a conductive polymer, polypyrrole, in order to increase their conductivity. With these modified nanowires, it was possible to electrochemically detect the important neurotransmitter dopamine. Additionally, the polymer-modified nanowires were used as a cellular platform for the growth of

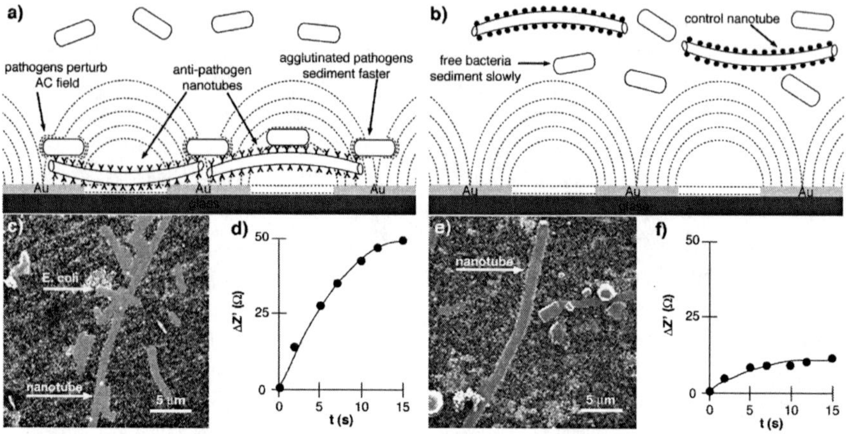

Figure 3.6: Label-free detection of pathogens with antibody-modified peptide nanotubes. a) Bacteria agglutinated by the interaction with antibodies on peptide nanotubes sedimented quickly and the increased number of insulating cells on the transducer via the sedimentation raised the impedance at 316 kHz. b) Control nanotubes modified with rabbit IgG that do not interact specifically with *E. coli* and did not sediment these cells fast enough to generate the impedance signal. c) SEM image of E. coli agglutinated by anti-*E. coli* nanotubes. d) An increase of the real part of the impedance (Z0) at 316 KHz with time due to the fast sedimentation of bacteria–nanotube complexes in (c) on the transducer. e) SEM image of the control nanotube modified with rabbit IgG that does not interact with E. coli. f) A significantly smaller Z0 change as compared to (d) when E. coli cells were mixed with control nanotubes in (e). de la Rica et al. 2010. Peptide-nanotube biochips for label-free detection of multiple pathogens. Small 6. 1092–1095. Copyright Wiley-VCH Verlag GmbH & Co. KGaA. Reproduced with permission.

HeLa and PC-12 cells, displaying the option to join these two features to develop a combined cellular platform/sensor device where the cells releasing the biomedical compounds to be detected are in intimate contact with the biorecognition element (Sasso et al. 2011).

Proteins such as the porin MspA are able to self-assemble into nanopores that can potentially be used as sensors. Porin MspA was used by Butler et al. for the development of a nanopore for nucleic acid analysis displaying potential for single-molecule detection. Single molecules of ssDNA were electrophoretically driven through the pore and could be electronically detected (Butler et al. 2008). Figure 3.7 shows a schematic diagram of this experiment and the representative ionic current signal observed.

Optical Diagnostic Devices

The functionalization of self-assembled nanostructures with photoluminescence (PL) compounds leads to new alternatives for the use of these supramolecular structures in the development of sensing

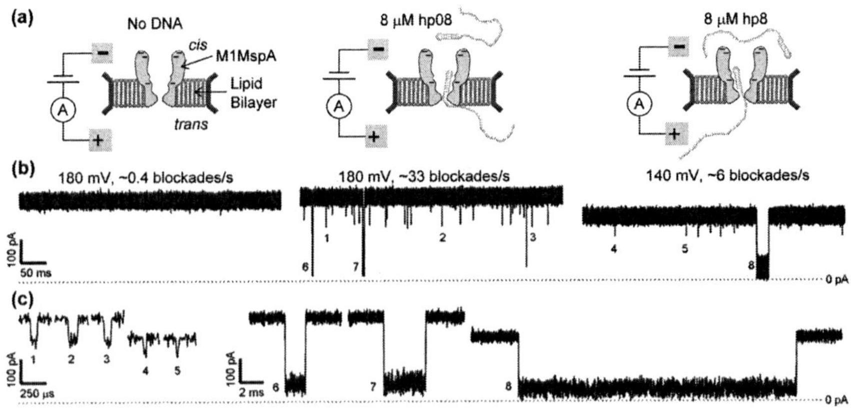

Figure 3.7: Detection of ssDNA hairpin constructs with M1MspA. (A) Schematic diagram of the experiments. (*B*) Representative ionic current signal observed for M1MspA in the absence of DNA and in the presence of 8-M hp08 hairpin DNA at 180 and 140 mV. (*C*) Numbered blockades from traces in *B* shown at expanded time scales.

devices based, in this case, on changes in light intensity. Ryu et al. functionalized diphenylalanine nanotubes with photosensitizers/lanthanice ions such as salicylic acid/Tb or 1,10-phenanthroline/Eu. The incorporation of lanthanide complexes into peptide nanotubes enhanced the photoluminescence of the Tb and Eu ions (Ryu et al. 2009). Based on this system, Kim et al. developed a sensor for the selective detection of paraoxon, which is a neurotoxin that inhibits the activity of acetylcholine esterase in the brain, thus leading to the termination of synaptic transmission (Kim et al. 2011a). In the presence of paraoxon, the photoluminescence of the functionalized peptide nanotubes readily quenched all the emission peaks within a few seconds. Figure 3.8 gives the suggested mechanism for the selective detection of the neurotoxin using photoluminescence self-assembled peptide nanotubes.

Yan et al. reported on the use of a 9-fluorenylmethoxycarbonyl (Fmoc)-modified dipeptide nanofibrils that entangle further to form gels to encapsulate quantum dots. In this way, the encapsulated quantum dots were protected from oxidation, which improved their stability (Yan et al. 2008). A similar approach as the one presented by Yan and co-workers was used for the development of a biosensing platform through the encapsulation of glucose oxidase and horseradish peroxidase as well as CdTe and CdSe quantum dots in an Fmoc-diphenylalanine hydrogel (Kim et al. 2011b). This functionalized peptide hydrogel was employed to detect glucose and toxic phenolic compounds by following the photoluminiscent quenching of the hybridized quantum dots. The obtained Michelis-Menten constant was found to be 3.12 mM (GOx for glucose) and 0.82 mM (HRP for hydroquinone) suggesting that the developed platform

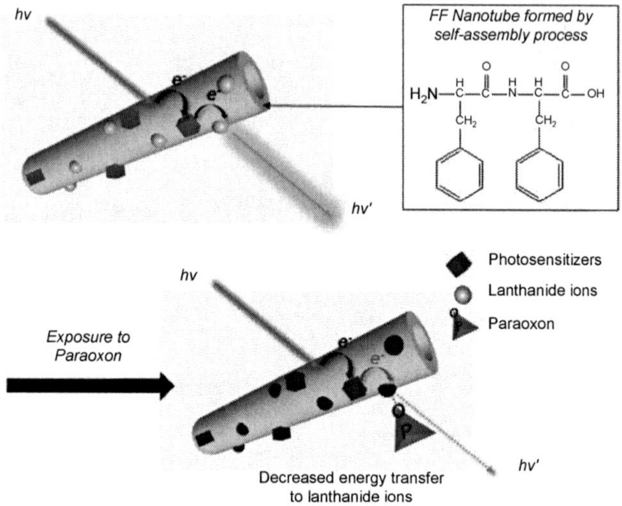

Figure 3.8: Suggested mechanism for the selective detection of paraoxon using photoluminescent FF nanotubes. FF nanotubes that are hybridized with lanthanide complexes exhibit an intense PL due to the transfer of cascaded energy from FF to lanthanide ions via photosensitizers. Upon the exposure to paraoxon, their PL could be readily quenched due to the inhibition of the cascaded-energy transfer. (Kim et al. 2011. Selective detection of neurotoxin by photoluminescent peptide nanotubes. Small 6. 718–722. Copyright Wiley-VCH Verlag GmbH & Co. KGaA. Reproduced with permission.)

Color image of this figure appears in the color plate section at the end of the book.

was an alternative optical biosensing platform. Figure 3.9 shows the key steps for the development of the photoluminiscent self-assembled peptide hydrogel and its application to the enzymatic detection of analytes.

A virus assay using antibody-functionalized peptide nanotubes was developed by MacCuspie and collaborators. In their work, viruses such as herpes simplex virus type 2, adenovirus, vaccinia and influenza type B were detected in attomolar amounts by monitoring changes in fluorescence and light scattering intensities related to the aggregation of dye-loaded antibody nanotubes around the viruses (MacCuspie et al. 2008). Nanotubes with average dimensions of 120 nm (diameter) x 1 μm (length) were prepared using bis(N-α-amidoglycine)-1,7-heptane dicarboxylate. Once formed, the tubes were functionalized with polyclonal antibodies for the virus detection.

Finally, Table 3.2 lists various applications of nanostructures of self-assembled peptides and proteins for the detection of compounds of biomedical relevance. In this table, the self-assembled nanostructure, the measuring technique, and the detected compound are introduced.

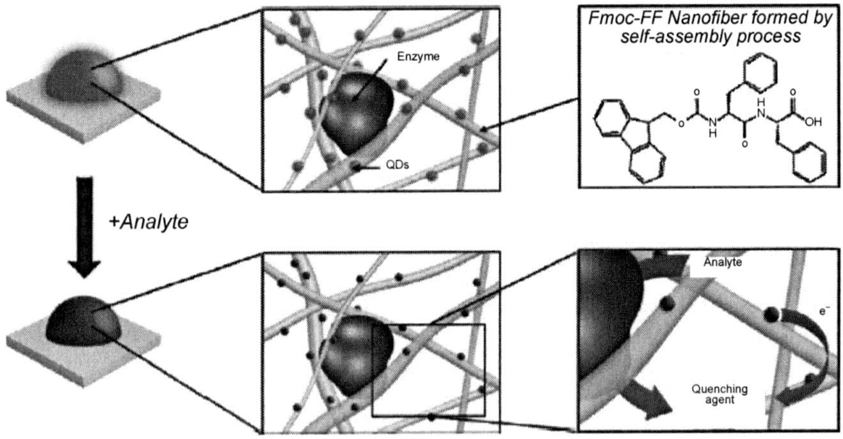

Figure 3.9: An illustrative description of the development of a photoluminescent peptide hydrogel through the self-assembly of Fmoc-FF building blocks and their PL quenching associated with the enzymatic detection of analytes. (Reprinted from Biosensors & Bioelectronics, 26, Kim et al. 2011. Self-assembled, photoluminiscent peptide hydrogel as a versatile platform for enzyme-based optical biosensors, 1860–1865, Copyright 2011, with permission from Elsevier.)

Color image of this figure appears in the color plate section at the end of the book.

Table 3.2: Application of self-assembled peptide and protein nanostructures for the detection of compounds of biomedical relevance.

Self-assembled nanostructure	Technique	Detected compound	Reference
Nanotube	Amperometric	Detection of glucose, ethanol and hydrogen peroxide	(Yemini et al. 2005a, Yemini et al. 2005b)
Nanowires	Amperometric	Glucose	(Yang et al. 2009)
Nanotubes	Amperometric	Hydrogen peroxide	(Cipriano et al. 2010)
Nanotubes	Voltammetry	*Eschericia coli* O157:H7	(Cho et al. 2008)
Nanofibrils	Voltammetry	Copper	(Viguier et al. 2011)
Nanotubes	Impedance	Herpes simplex virus type 2	(de la Rica et al. 2008)
Nanotubes	Impedance	*Eschericia coli* and *S. typhi* cells	(de la Rica et al. 2010)
Nanowires	Amperometry	Dopamine	(Sasso et al. 2011)
Nanopore	Amperometry	ssDNA hairpins	(Butler et al. 2008)
Nanotubes	Photoluminiscent	Paraoxon	(Kim et al. 2011)
Nanofibrils hydrogel	Photoluminiscent	Glucose and hydroquinones	(Kim et al. 2011)
Nanotubes	Fluorescence	Herpes simplex virus type 2, adenovirus, vaccinia and influenza type B	(MacCuspie et al. 2008)
Nanowire	Optical adsorption	*Yersinia pestis*	(Men et al. 2010)

CONCLUSION

As presented in this chapter, self-assembled peptide and self-assembled protein nanostructures constitute a very interesting alternative for the development of diagnostic devices for application in the biomedical field. They offer several advantages in terms of fabrication, functionalization and biocompatibility. Additionally they can be synthesized in different structural shapes such as tubes, wires, particles, tapes or hydrogels. Despite these advantages, challenges related to size and shape control during the synthesis process and the manipulation and integration during the integration with transducers need to be confronted. Some of the techniques and methods available to overcome these challenges were presented and examples of their use were discussed.

Finally, representative applications of these biological nanostructures in the development of diagnostic platforms were presented. Examples on the electrochemical, electrical and optical detection of compounds of biomedical relevance supporting the key role these types of nanostructures in nanomedicine were discussed and tabulated.

New advances in manipulation techniques and surface chemistry will accelerate and increase the use of these biological nanostructures as an alternative to traditionally used nanomaterials in biomedical applications.

REFERENCES

Ahmed, M.U. and M.M. Hossain and E. Tamiya. 2008. Electrochemical biosensors for medical and food applications. Electroanalysis 20: 616–26.

Butler, T.Z. and M. Pavlenok, I.M. Derrington, M. Niederweis and J.H.Gundlach. 2008. Single-molecule DNA detection with an engineered MspA protein nanopore. Proceedings of the National Academy of Sciences of the United States of America 105: 20647–52.

Castillo, J. and S. Tanzi, M. Dimaki and W.Svendsen. 2008. Manipulation of self-assembly amyloid peptide nanotubes by dielectrophoresis. Electrophoresis 29: 5026–32.

Castillo, J. and M. Dimaki and W.E. Svendsen. 2009. Manipulation of biological samples using micro and nano techniques. Integrative Biology 1: 30–42.

Castillo, J. and W.E. Svendsen and M. Dimaki. 2011. Micro and Nano Techniques for the Handling of Biological Samples. J. Castillo, W.E. Svendsen, M. Dimaki, New York.

Castillo-León et al. 2011 with Castillo-León, J. and R. Rodriguez-Trujillo, S. Gauthier, A.C.Ø. Jensen and W.E. Svendsen. 2011. Micro-"factory" for self-assembled peptide nanostructures. Microelectronic Engineering 88: 1685–1688.

Cho, E.C. and J.W. Choi, M.Y. Lee and K.K. Koo. 2008. Fabrication of an electrochemical immunosensor with self-assembled peptide nanotubes. Colloids and Surfaces a-Physicochemical and Engineering Aspects 313: 95–99.
Cipriano, T.C. and P.M. Takahashi and D. de Lima et al. 2010. Spatial organization of peptide nanotubes for electrochemical devices. Journal of Materials Science 45: 5101–08.
Daamen, W.F. and P.J. Geutjes and H.T.B. van Moerkerk et al. 2007. Lyophilisomes: A new type of (bio)capsule. Advanced Materials 19: 673–77.
de la Rica, R. and H. Matsui. 2010. Applications of peptide and protein-based materials in bionanotechnology. Chemical Society Reviews 39: 3499–509.
de la Rica, R. and E. Mendoza, L.M. Lechuga and H. Matsui. 2008. Label-Free Pathogen Detection with Sensor Chips Assembled from Peptide Nanotubes. Angewandte Chemie-International edn. 47: 9752–55.
de la Rica, R. and C. Pejoux, C. Fernandez-Sanchez, A. Baldi and H.Matsui. 2010. Peptide-Nanotube Biochips for Label-Free Detection of Multiple Pathogens. Small 6: 1092–95.
Frauenrath, H. and E. Jahnke. 2008. A general concept for the preparation of hierarchically structured pi-conjugated polymers. Chemistry-a European Journal 14: 2942–55.
Gao, X.Y. and H. Matsui. 2005. Peptide-based nanotubes and their applications in bionanotechnology. Advanced Materials 17: 2037–50.
Ghadiri, M.R. and J.R. Granja R.A. Milligan, D.E. McRee and N. Khazanovich. 1993. Self-assembling organic nanotubes based on a cyclic peptide architecture. Nature 366: 324–27.
Gras, S.L. and J.K. Rudy. Surface- and Solution-Based Assembly of Amyloid Fibrils for Biomedical and Nanotechnology Applications. pp.161–209. *In:* 2009. Advances in Chemical Engineering. Academic Press,
Hauser, C.A.E. and S.G. Zhang. 2010. Designer self-assembling peptide nanofiber biological materials. Chemical Society Reviews 39: 2780–90.
He, B. and T.J. Morrow and C.D. Keating. 2008. Nanowire sensors for multiplexed detection of biomolecules. Current Opinion in Chemical Biology 12: 522–28.
Holmes, T.C. and S. de Lacalle, X. Su, G.S. Liu, A. Rich and S.G. Zhang. 2000. Extensive neurite outgrowth and active synapse formation on self-assembling peptide scaffolds. Proceedings of the National Academy of Sciences of the United States of America 97: 6728–33.
Huster, D. and S. Lutsenko. 2007. Wilson disease: not just a copper disorder. Analysis of a Wilson disease model demonstrates the link between copper and lipid metabolism. Mol Biosyst 3: 816–24.
Kim, T. and Y.M. Huh, S. Haam and K. Lee. 2010. Activatable nanomaterials at the forefront of biomedical sciences. Journal of Materials Chemistry 20: 8194–206.
Kim, J.H. and J. Ryu and C.B. Park. 2011. Selective Detection of Neurotoxin by Photoluminescent Peptide Nanotubes. Small 7: 718–22.
Kim, J.H. and S.Y. Lim, D.H. Nam, J. Ryu, S.H. Ku and C.B. Park. 2011. Self-assembled, photoluminescent peptide hydrogel as a versatile platform for enzyme-based optical biosensors. Biosensors & Bioelectronics 26: 1860–65.

Kumara, M.T. and N. Srividya, S. Muralidharan and B.C. Tripp. 2006. Bioengineered flagella protein nanotubes with cysteine loops: Self-assembly and manipulation in an optical trap. Nano Letters 6: 2121–29.

MacCuspie, R.I. and I.A. Banerjee and C. Pejoux et al. 2008. Virus assay using antibody-functionalized peptide nanotubes. Soft Matter 4: 833–39.

Matsui, H. and B. Gologan, S. Pan and G.E. Douberly. 2001. Controlled immobilization of peptide nanotube-templated metallic wires on Au surfaces. European Physical Journal D 16: 403–06.

Men, D. and Z.P. Zhang and Y.C. Guo et al. 2010. An auto-biotinylated bifunctional protein nanowire for ultra-sensitive molecular biosensing. Biosensors & Bioelectronics 26: 1137–41.

Miranda, F.F. and K.Iwasaki and S. Akashi et al. 2009. A Self-Assembled Protein Nanotube with High Aspect Ratio. Small 5: 2077–84.

Perez-Lopez, B. and A. Merkoci. 2011. Nanoparticles for the development of improved (bio)sensing systems. Analytical and Bioanalytical Chemistry 399: 1577–90.

Porrata, P. and E. Goun and H. Matsui. 2002. Size-controlled self-assembly of peptide nanotubes using polycarbonate membranes as templates. Chemistry of Materials 14: 4378–81.

Qureshi, A. and W.P. Kang, J.L. Davidson and Y. Gurbuz. 2009. Review on carbon-derived, solid-state, micro and nano sensors for electrochemical sensing applications. Diamond and Related Materials 18: 1401–20.

Ramgir, N.S. and Y. Yang and M. Zacharias. 2010. Nanowire-Based Sensors. Small 6: 1705–22.

Reches, M. and E. Gazit. 2004. Formation of closed-cage nanostructures by self-assembly of aromatic dipeptides. Nano Letters 4: 581–85.

Reches, M. and E. Gazit. 2006. Designed aromatic homo-dipeptides: formation of ordered nanostructures and potential nanotechnological applications. Physical Biology 3: S10–S19.

Reches, M. and E. Gazit. 2006. Controlled patterning of aligned self-assembled peptide nanotubes. Nature Nanotechnology 1: 195–200.

Reches, M. and E. Gazit. 2007. Biological and chemical decoration of peptide nanostructures via biotin-avidin interactions. J Nanosci Nanotechnol 7: 2239–45.

Ryu, J. and S.Y. Lim and C.B. Park. 2009. Photoluminescent Peptide Nanotubles. Advanced Materials 21: 1577–81.

Sasso, L. and I. Vedarethinam, J. Emnéus, W.E. Svendsen and J. Castillo-León. 2011. Self-Assembled Diphenylalanine Nanowires for Cellular Studies and Sensor Applications J Nanosci Nanotechnol (in press).

Scheibel, T. and R. Parthasarathy, G. Sawicki, X.M. Lin, H. Jaeger and S.L. Lindquist. 2003. Conducting nanowires built by controlled self-assembly of amyloid fibers and selective metal deposition. Proceedings of the National Academy of Sciences of the United States of America 100: 4527–32.

Sedman, V.L. and S. Allen, X.Y. Chen, C.J. Roberts and S.J.B. Tendler. 2009. Thermomechanical Manipulation of Aromatic Peptide Nanotubes. Langmuir 25: 7256–59.

Siangproh, W. and W. Dungchai, P. Rattanarat and O. Chailapakul. 2011. Nanoparticle-based electrochemical detection in conventional and miniaturized systems and their bioanalytical applications: A review. Analytica Chimica Acta 690: 10–25.

Smith, A.M. and R.J. Williams and C. Tang et al. 2008. Fmoc-Diphenylalanine self assembles to a hydrogel via a novel architecture based on pi-pi interlocked beta-sheets. Advanced Materials 20: 37–41.

Toksoz, S. and H. Acar and M.O. Guler. 2010. Self-assembled one-dimensional soft nanostructures. Soft Matter 6: 5839–49.

Usui, K. and T. Maki and F. Ito et al. 2009. Nanoscale elongating control of the self-assembled protein filament with the cysteine-introduced building blocks. Protein Science 18: 960–69.

Viguier, B. and K. Zór and E. Kasotakis et al. 2011. Development of an Electrochemical Metal-Ion Biosensor Using Self-Assembled Peptide Nanofibrils. ACS Applied Materials & Interfaces. (in press).

Wang, J. 2005. Carbon-nanotube based electrochemical biosensors: A review. Electroanalysis 17: 7–14.

Yan, X.H. and P.L. Zhu and J.B. Li. 2010. Self-assembly and application of diphenylalanine-based nanostructures. Chemical Society Reviews 39: 1877–90.

Yang, H. and S.Y. Fung, M. Pritzker and P. Chen. 2009. Ionic-Complementary Peptide Matrix for Enzyme Immobilization and Biomolecular Sensing. Langmuir 25: 7773–77.

Yemini, M. and M. Reches, E. Gazit and J. Rishpon. 2005. Peptide nanotube-modified electrodes for enzyme-biosensor applications. Analytical Chemistry 77: 5155–59.

Yemini, M. and M. Reches, J. Rishpon and E. Gazit. 2005. Novel electrochemical biosensing platform using self-assembled peptide nanotubes. Nano Letters 5: 183–86.

Zhang, S.G. and T. Holmes, C. Lockshin and A. Rich. 1993. Spontaneous assembly of a self-complementary oligopeptide to form a stable macroscopic membrane. Proceedings of the National Academy of Sciences of the United States of America 90: 3334–38.

Zhao, Z. and H. Matsui. 2007. Accurate immobilization of antibody-functionalized peptide nanotubes on protein-patterned Arrays by optimizing their ligand-receptor interactions. Small 3: 1390–93.

Zhao, Z.Y. and P.A. Banerjee. and H. Matsui. 2005. Simultaneous targeted immobilization of anti-human IgG-coated nanotubes and anti-mouse IgG-coated nanotubes on the complementary antigen-patterned surfaces via biological molecular recognition. Journal of the American Chemical Society 127: 8930–31.

Zimenkov, Y. and S.N. Dublin and R. Ni et al. 2006. Rational design of a reversible pH-responsive switch for peptide self-assembly. Journal of the American Chemical Society 128: 6770–71.

4

Electrochemical Sensor Systems for Medicine

V.V. Shumyantseva,[1,a,*] *E.V. Suprun,*[1] *T.V. Bulko,*[1] *Ya. M. Chalenko*[1] and *A.I. Archakov*[1]

INTRODUCTION

One of the major challenges of engineering (technical) nanotechnology lies in coupling nano-sized bio-objects with measuring analytical devices. Preparation of hybrid bio-organic nanocomposite materials provides a 'bridge' for such coupling. Functional significance of hemoproteins in the body (in living organisms) is exceedingly high. Participating in various metabolic processes within living creatures, hemoproteins play an important role in maintenance of life on earth. Hemoglobin and myoglobin perform binding and carrying of molecular oxygen. Cytochrome c acts by transferring electrons in the mitochondrial respiratory chain. Cytochrome c oxydase catalyzes terminal oxidation with participation of molecular oxygen, while catalase is responsible for peroxide itself. NO-synthase brings about the synthesis of nitric oxide from L-arginine. Cytochromes P450 are involved into hydroxylation of organic molecules (Scheme 4.1).

The level of hemoproteins in organs, tissues, and body fluids is an important parameter of normal organism functioning in general and therefore development of various tests systems for express assay of these

[1]Institute of Biomedical Chemistry, Russian Academy of Medical Sciences, Pogodinskaya St. 10, Moscow 119121, Russia.
[a]E-mail: viktoria.shumyantseva@ibmc.msk.ru
*Corresponding author

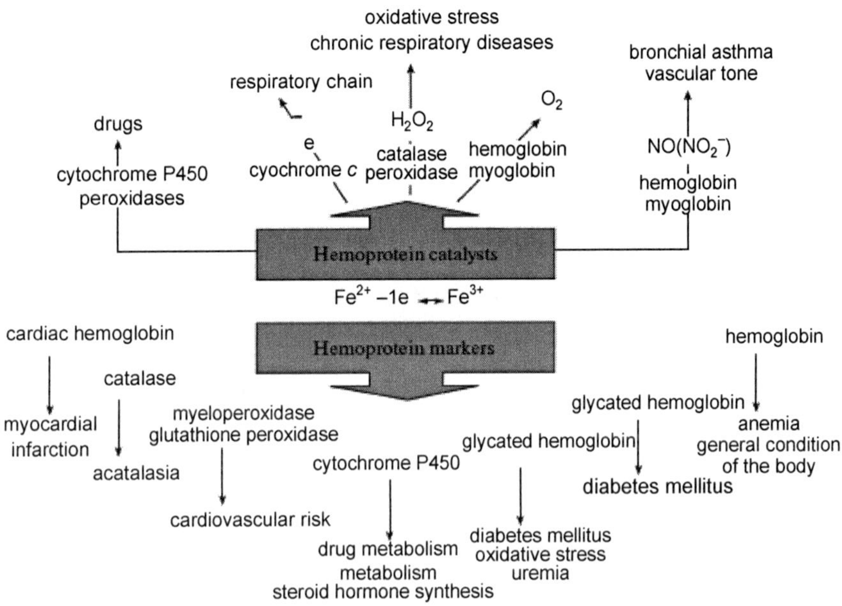

Scheme 4.1: Clinical importance of hemoproteins.

proteins is very important. For example, assay of blood hemoglobin is a routine test in medical practice. Modified hemoglobins, glutathionyl-hemoglobin, glycohemoglobin, are markers in diabetic subjects, uremic patients, and in patients with Friedrech's ataxia (Dalle-Donne et al. 2006). Myocardial myoglobin is the earliest cardiac marker among all markers that appear in blood (McDonnel et al. 2009).

Redox activity of a catalytic site of hemoproteins and a wide range of reactions catalyzed by them make hemoproteins important instruments in the development of electrochemical biosensors. Construction of platforms for enzyme immobilization and signal detection involves progressive nanotechnologies and nanomaterials, including nanofibers, carbon nanotubes, modification of a detector surface with gold nanoparticles, polymer films, and various polymer membranes (Xiao and Li 2008).

ELECTROCHEMICAL INVESTIGATIONS OF CYTOCHROMES P450

P450s enzymes represent a superfamily of b-type hemeproteins with catalytic activity towards two substrates: oxygen and organic substance. Cytochromes P450 play an important role in detoxication of bioactive compounds and hydrophobic xenobiotics both coming from outside (medicines, drugs, food supplements, environment pollutants) and being formed inside cells (cholesterol, saturated and unsaturated fatty acids, steroids, prostaglandins and others) in living organisms (Archakov and Bachmanova 1990, Lewis 2001, Ortiz de Montellano 2004). Prediction of metabolic transformation of physiologically active substances and investigation of possible metabolic pathways are conducted by use of computer-aided methods *in silico* (Balakin and Ivanenkov 2005), or experimentally with enzymatic systems *in vitro* (Cirspi 1999).

Since P450-based test systems are in great demand, various methods for high-throughput screening of substrates and inhibitors are being actively developed (Persson et al. 2006, Turpeinen et al. 2005). The majority of proposed methods are based (a) on measuring, by direct techniques (LC/MS, HPLC), the activity of human liver microsomes or the activity of recombinant cytochromes P450, each expressed together with reductase (baculosomes), against studied substrates; or (b) on measuring the activity of cytochromes P450 by the fluorescent probe method. The analysis of cassette incubation of probe substrates ('the cocktail') approach or n-in-one technique is also employed. Spectral optical methods can only provide information about the type of binding although their efficiency is high enough.

Investigation of catalytic activity of isolated cytochromes from the P450 superfamily requires the obligatory presence of redox partners and electron donors (NADPH) (Lewis 2001). However, redox partners are not compelled by electrochemical reduction of P450 family hemoproteins, so the catalytic system is essentially simplified (Shumyantseva et al. 2005). The electrochemical approach is especially important in the case of unknown physiological partners (e.g., CYP51 MT, systematic name CYP51b1) (Bellamine 1999, Pikuleva 2006, Hlavica 2009). Electrochemical systems execute a dual function: substitute partner proteins and serve as a source of electrons for redox enzymes.

From the chemical viewpoint, this diversity of functions is based on oxidative-reductive properties of the P450s' heme iron. From the electrochemical viewpoint, the most characteristic feature of hemeproteins' is their ability to perform direct electron transfer from the heme iron to the electrode surface, i.e., the appearance on the voltammogram of a pair of peaks corresponding to the oxidation and reduction processes. The electrode reaction of the heme may be described by the equation:

$$\mathrm{Fe(III)} + e^- + H^+ \leftrightarrow \mathrm{Fe(II)}$$

The relevance of such an approach is apparent: indeed, cytochromes P450-based enzyme electrodes may be used as biosensors in personalized medicine, high-throughput screening and drug interference studies. From this viewpoint electrochemical systems based on recombinant forms of P450s are most promising because they enable in standardizing the analysis format (Estabrook et al. 1996, Bistolas et al. 2005, Shumyantseva et al. 2005, Udit et al. 2005, Shumyantseva et al. 2011b).

Studies aimed at the search of new drugs, estimation of their toxicity and drug-drug interaction have shown that cytochromes P450 are the most significant preparations for use in practical clinical medicine. Over 270 gene families, encoding cytochromes P450 (CYP genes), have been identified. Clinical significance of cytochromes P450 is connected with their functions in the metabolism of exogenous and endogenous compounds. Cytochromes P450 play an important role in the living organism on detoxification of biologically active compounds and hydrophobic xenobiotics—either penetrating externally or formed in the cell (Nebert and Russell 2002). Endogenous substrates for cytochromes P450 are saturated and unsaturated fatty acids, eucozanoids, steroidic hormones, vitamin D, vitamin E, retinoids, bile acids, prostaglandins, uroporphyrogens, arachidonic acid (http://cpd.ibmh.msk.su; Knowledge based on P450s). Exogenous substrates for human cytochromes P450 are medicinal drugs, externally penetrating compounds (pesticides, herbicides and plant components). Xenobiotic metabolism proceeds with participation of human cytochromes' families CYP1, CYP2, CYP3 and, to a lesser degree, CYP4. For instance P450 1A2 metabolizes 106 substrates of which 72 substances are medicinal preparations (caffeine, paracetamol, phenacetin) (Lisitsa et al. 2001). Five major cytochrome P450 forms out of 40 human cytochromes P450 are responsible for metabolism of over 87% of medicinal drugs. These are: CYP3A4, CYP2D6, CYP1F2, CYP2C9, CYP2C19 (Smith et al. 2007).

Electrochemical approaches are important for studying enzyme-substrate interaction due to their high sensitivity (Shumyantseva et al. 2004, Bistolas et al. 2005, Udit and Gray 2005, Joseph et al. 2003, Sadegni S. et al. 2011). A special feature of P450-based electrochemical sensors is the usage of nanostructured electrodes for improveming the sensitivity of analyses. Amperometric CYP2B4-based electrochemical biosensor was proposed for determining phenobarbital as a substrate of cytochrome P450 (detection limit 0.289 μM). Phenobarbital is a medicinal drug for use in treatment of epilepsy (Alonso-Lomillo et al. 2008). CYP2B6 was incorporated into chitosan modified colloidal gold nanoparticles films. CYP2B6-electrodes were used for the investigation of electrocatalytic behavior of CYP2B6 towards lidocaine, bupropion, cyclophosphamide (Liu et al. 2008). Direct electrochemistry of P450 6A1 was studied on edge-plane graphite electrode and the catalytic activity to aldrin has been demonstrated with cyclic voltammetry (Zhang et al. 2010). Attempts to develop a potentiometric analysis of enzyme-substrate interaction proved to have little effect since no direct correlation between the shifts of reduction potential with and without substrate were established. Besides, such an approach was ineffective in the search of cytochromes' P450 inhibitors. Inhibitors of cytochromes P450 may be potential medical preparations by influencing one or another metabolic pathway and by lowering the activity of one or another hemoprotein, responsible for drug metabolism. Cytochromes P450 also serve as a target object for development of new anti-cancer drugs in the case of hormone dependent cancer diseases (Bruno and Njar 2007). Potential inhibitors of CYP19 (aromatase) are used in treatment of breast cancer as estrogen dependent tumors. Inhibitors of CYP17 (17-α-hydroxylase) are finding application in treatment of androgen dependent prostate cancer.

To develop the algorithm for the search of potential substrates and inhibitors of cytochromes P450 we tested various electrochemical methods. Gold nanoparticles, stabilized by the synthetic membrane-like surfactant didodecyldimethylammonium bromide (DDAB), provide effective electron transport between graphite electrode and the cytochrome P450 heme. Synthetic lipid membrane with colloidal gold, DDAB/Au, contains a sufficient amount of water to maintain the hemeprotein's structure and to secure fixation of P450 enzymes on graphite screen printed electrodes. Cytochromes P450 2B4, 1A2, 3A4, 11A1 (P450scc), P450 51b1 (CYP51), CYP17A1 were studied in the presence of substrates and/or inhibitors of these forms (Shumyantseva et al. 2007b, Shumyantseva et al. 2010b, Shumyantseva et al. 2011a, Carrara et al. 2008).

To investigate electro analytic characteristics, current-voltage electrode responses were used; they were registered by use of cyclic voltammetry and voltammetric analysis (square wave and differential pulse voltammetry). Substrates for appropriate forms of cytochromes P450 cause substantial enhancement of catalytic current at controlled potential while inhibitors do not change or even lower the current responses.

For studying a number of enzymes and/or substrates 96-well electrochemical, readers were referred to (Alderon biosciences, http://www/alderonbiosciences.com). It is to be noted that in studies under the classical scheme by the square-wave voltammetric method or by multichannel micropotentiostat, the close values of current changes were obtained. The increase of catalytic current in the presence of substrate for appropriate form were registered, inhibition of the current on addition of inhibitors and inhibition of catalytic current in the case of initially added inhibitor and subsequently added substrate.

Thus based on the results of amperometry, cyclic voltammetry, differential pulse voltammetry DPV, square-wave voltammetry SWV, and intermittent pulse amperometry (IPA), it is possible to conduct a search and investigate kinetic parameters for potential substrates and inhibitors of cytochromes P450. The proposed electrochemical approach is a sort of a bio bar code for determination of the cytochromes P450 substrate/inhibitor competence.

Electrochemical methods may be coupled with tandem liquid chromatography-mass-spectrometry (LC/MS/MS). Analysis of electrochemical reaction products by use of LC/MS has shown the presence of the same substances as were revealed in the case of enzymatic reaction in the presence of cofactors and partner proteins (Shumyantseva et al. 2007a, Liu 2008).

Cytochrome P450 3A4 plays the major functional role as compared to other P450s: it metabolizes 225 substrates of which 191 are therapeutic drugs; among the 97 inhibitors of this enzyme, 87 compounds are therapeutic drugs (Lisitsa et al. 2001). Cytochrome P450 3A4 is involved in the metabolism of 34% of therapeutic drugs which are now in use (Lewis 2001). Analysis of interaction of cytochrome P450 3A4 with testosterone by differential pulse voltammetry (DPV) reveals the presence of catalytic current upon enzyme-substrate interactions, (see Fig. 4.1 and Fig. 4.2, experiments 14–17). The ratio of maximal amplitudes of reductive currents in the presence of testosterone and without substrate may be expressed as $I_{DP}(O_2 + Ts)/I_{DP}(O_2) = 1.7$; the inhibitor ketoconazole (36 mM) lowers this ratio to 1.3.

Figure 4.2 presents maximal peak currents of SW voltammograms, corrected by the baseline, for enzyme electrodes in the presence of substrates and/or inhibitors: maximal amplitude of the current before the addition of the substrate benzphetamine (1) and after (2) the addition of this substrate; after the addition of camphor (3)—which is not a

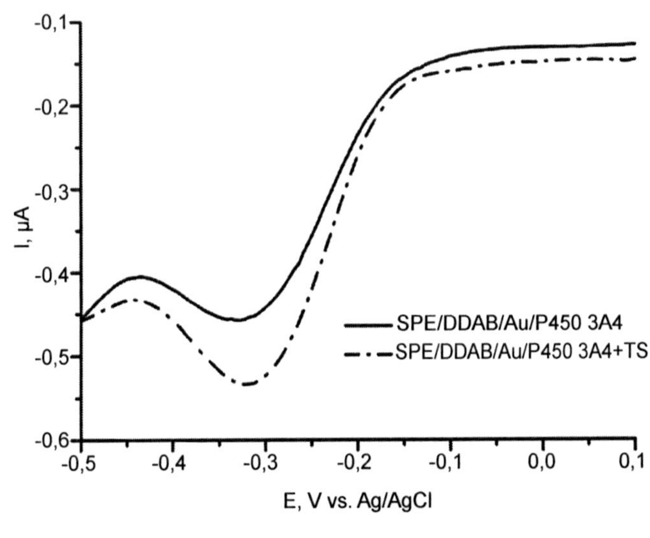

Figure 4.1: Reductive DPV of screen-printed DDAB/Au/P450 3A4 electrode before (-) and after the addition of 200 µM testosterone (TS) (-·-). Electrolyte volume is 1 ml 100 mM potassium phosphate buffer plus 50 mM NaCl, pH 7.4.

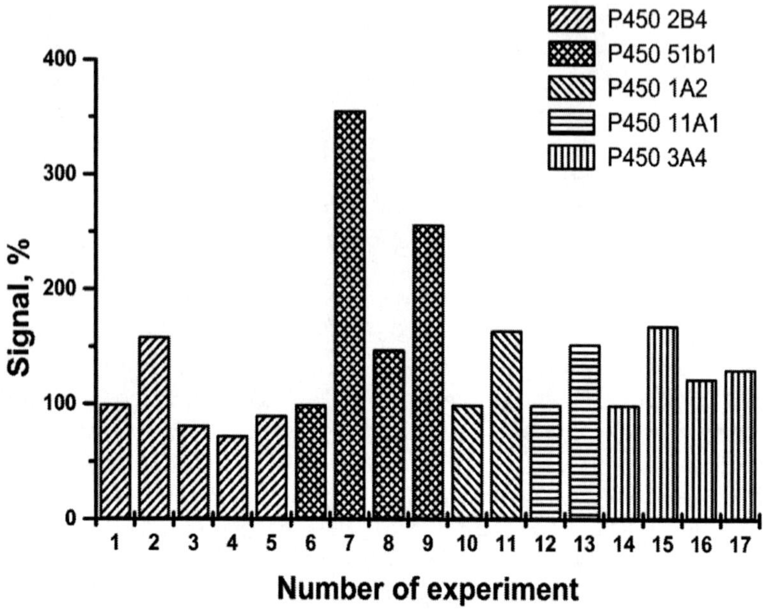

Figure 4.2: Peak intensity of reductive SWV of screen-printed electrodes in aerobic buffer (with baseline correction):

1) DDAB/Au/P450 2B4
2) DDAB/Au/P450 2B4 + 250 μM benzphetamine (Bz)
3) DDAB/Au/P450 2B4 + 25 μM camphor
4) DDAB/Au/P450 2B4 + 1 mM metyrapone
5) DDAB/Au/P450 2B4 + 1 mM metyrapone + 250 μM benzphetamine (Bz)
6) DDAB/Au/P450 51b1
7) DDAB/Au/P450 51b1 +10 μM lanosterol
8) DDAB/Au/P450 51b1 + 36 μM ketoconazole,
9) DDAB/Au/P450 51b1 +36 μM ketoconazole + 10 μM lanosterol
10) DDAB/Au/P450 1A2
11) DDAB/Au/P450 1A2 +5 μM 7-ethoxyresorufin
12) DDAB/Au/P450 11A1
13) DDAB/Au/P450 11A1 + 200 μM cholesterol
14) DDAB/Au/P450 3A4
15) DDAB/Au/P450 3A4 + 200 μM testosterone
16) DDAB/Au/P450 3A4 + 36 μM ketoconazole
17) DDAB/Au/P450 3A4++ 36 μM ketoconazole + 200 μM testosterone

substrate for P450 2B4; after the addition of the inhibitor metyrapone (4) and, lastly, after the addition of the substrate benzphetamine (5). The amplitude of catalytic current of SW voltammogram was increased upon benzphetamine addition up to 160% (2); in the presence of metyrapone a 20% decrease of the current amplitude occurs as compared to the basic value (4). The mechanism of inhibitory action of metyrapone is based on its binding to the heme iron and also on the lipophilic binding of metyrapone to the hydrophobic regions of the protein (Ortiz de Montellano 1995). The decrease of peak current is possibly explained by the competition of metyrapone with oxygen for binding to the heme iron and for substrate binding. Camphor does not cause the increase of catalytic current (3) because it is not metabolized by cytochrome P450 2B4 (Lewis 2001).

Similar results were obtained in our earlier studies of electro catalysis of sterol-14α-demethylase (CYP51 MT, P450 51b1) in the presence of the substrate lanosterol and the inhibitor ketoconazole (Fig. 4.3, experiments 6–9) (Shumyantseva et al. 2007a).

The electrochemical reduction of the recombinant form of human cytochrome P450 17A1 (CYP17A1) was investigated. Hemoprotein was immobilized on electrode modified with biocompatable nanocomposite material based on the membrane-like synthetic surfactant didodecyldimethylammonium bromide (DDAB) and gold nanoparticles. Analytical characteristics of DDAB/Au/CYP17A1 electrodes were investigated with cyclic voltammetry, square wave voltammetry, and differential pulse voltammetry. Analysis of electrochemical behavior of cytochrome P450 17A1 was conducted in the presence of substrate pregnenolone, inhibitor ketoconazole, and in the presence of synthetic derivatives of pregnenolone: acetylpregnenolone, cyclopregnenolone (4), and tetrabrompregnenolone (5). Ketoconazole, azole inhibitor of cytochromes P450, blocked catalytic current in the presence of substrate pregnenolone (1). Synthetic derivatives of pregnenolone: acetylpregnenolone, cyclopregnenolone, and tetrabrompregnenolone did not demonstrate substrate properties towards electrode/ CYP17A1 system. Acetylpregnenolone did not block catalytic activity towards pregnenolone, but cyclopregnenolone and tetrabrompregnenolone inhibited such activity. Electrochemical reduction of CYP17A1 may serve as an adequate substitution of the reconstituted system which requires additional redox partners—for the exhibition of catalytic activity of hemoproteins of the cytochrome P450 superfamily (Shumyantseva et al. 2011a).

Amperometric titration at controlled potential makes it possible to calculate the electrochemical Michaelis constant K_M for cytochrome P450 substrates (Table 4.1).

The proposed electrochemical approach can be useful for determination of the P450's substrate/inhibitor competence.

Stoichiometry of the electrocatalytic cycle of cytochrome P450 2B4 was studied in kinetic mode according to the bielectrode scheme (Rudakov et al. 2008). The bielectrode scheme registers oxygen consumption and hydrogen peroxide formation in kinetic mode during the reduction of hemoprotein. Based on the stoichiometry of P450 electrocatalysis, the adequacy of electrochemical reduction of P450 and P450-monooxygenase system was revealed. Stoichiometric ratios were determined for

Table 4.1: Electrochemical Michaelis constants K_M, calculated from the results of amperometric titration of DDAB/Au/P450 electrodes.

Electrode	K_M	
DDAB/Au/P450 2B4 +benzphetamine	13 μM	50 μM (Kanaeva et al. 1992)
DDAB/Au/P45011A1 + cholesterol	830 μM (upon titration of 14 mM standard cholesterol solution) 17 μM (upon titration of 10 mM cholesterol in ethanol)	K_d 4 μM (Lepesheva et al. 1999)
DDAB/Au/P45051b1 + lanosterol	30 μM	1 μM (Lepesheva and Waterman 2004)

electrocatalytic reduction of cytochrome P450 2B4 without substrate (ΔO_2:ΔH_2O_2 = 1:0.92) and in the presence of substrate benzphetamine (ΔO_2:ΔH_2O_2: formaldehyde = 1:0.15:0.8; formaldehyde is a product of benzphetamine N-demethylation). Based on the results of reaction rate measurement data and on the value of the stoichiometrical coefficient, it was concluded that the electrode-immobilized cytochrome P450 2B4 behaves in essentially the same way as does the mono-oxygenase microsomal system in solution.

The temperature dependence of redox potential $E^{0'}$ of cytochrome P450 2B4 allows the determination of thermodynamic parameters of the electron transfer process, in particular the standard entropy ($\Delta S^{0'}_{rc}$) and enthalpy $\Delta H^{0'}_{rc}$ changes associated with the reduction of the oxidized protein. Thermodynamics parameters of the electrochemical cycle of P450s are important factors that determine the driving force for the effective electron transfer between electrode and heme iron. Therefore, studies on the thermodynamics behavior of P450s help in gaining a better insight into the electron transfer kinetics and the properties of hemoprotein molecules. Dynamic electrochemistry (e.g., cyclic voltammetry) is a particularly useful tool for determination of electrochemical as well as thermodynamic characteristics of metalloenzymes (Liu et al. 2005). The dependence of redox potential of cytochrome P450 2B4 on temperature was determined in the temperature range of 5–30°C and thus the thermodynamic parameters of P450 2B4 were calculated (Shumyantseva et al. 2011b). Studies on the redox thermodynamics of P450s can provide a better understanding of the electron transfer mechanism and, hence, the improvement of sensitivity of electrochemical sensing systems. Electrochemically obtained partition of the enthalpic and entropic contributions to the redox potential of cytochrome P450 will be helpful to the elucidation of molecular mechanisms and factors that influence the protein's redox kinetics and electro catalysis. Our main goal was to determine the factorization of formal potential $E^{0'}$ into enthalpic and entropic contributions, each of which can be obtained through direct electrochemical experiments at variable temperatures, using screen-printed nanostructured electrodes and cytochrome P450 2B4. Measurements of $E^{0'}$ as a function of temperature allow the estimation of $\Delta S^{0'}_{rc}$ and $\Delta H^{0'}_{rc}$:

$$E^{0'} = \frac{-\Delta H^{0'}}{nF} + \frac{T\Delta S^{0'}}{nF}$$

Where n is the number of electrons, and F is Faraday's constant, 96485 C/mol.

The reaction entropy for the reduction of the oxidized cytochrome P450 2B4 is calculated from $\Delta S^{0'}_{rc} = nF\, dE^{0'}/dT$; thus $\Delta S^{0'}_{rc}$ was determined from the slope of $E^{0'}$ vs. temperature and corresponds to -115.7 $Jmol^{-1}K^{-1}$. The enthalpy change $\Delta H^{0'}_{rc}$ was obtained from the Gibbs-Helmholtz equation, namely as a negative slope of the $E^{0'}/T$ vs. $1/T$ plot and corresponds to -29.1 $kJ{\cdot}mol^{-1}$. The thermodynamics parameters for cytochrome P450 2B4 reduction,—$\Delta H^{0'}_{rc}$ and $\Delta S^{0'}_{rc}$—are both negative, which is typical for the hemeprotein's redox thermodynamics (He and Zhu 2006). The commonly accepted viewpoint is that the enthalpy loss is due to the predominance of the reduced heme of cytochrome P450 2B4 over its oxidized ferri-form and, very likely, due to the limited accessibility of the solvent in the reduced state. The entropic contribution and entropy loss upon reduction are accounted for conformational variations during the electrochemical cycle.

CYPs have an alternative route for mono-oxygenation via the 'peroxide shunt' in the presence of hydrogen peroxide (Hlavica 2009). The mechanism of this reaction presupposes heme-peroxide complex formation followed by insertion of oxygen atom into substrate molecule:

Fe(III) +H-O-O-H + RH → Fe(III)-O-OH (RH) → Fe(V)=O(RH) → Fe(III)+ROH

In this case, CYPs do not need electrons for heme iron reduction. Catalytic activity of peroxide-dependent reactions of cytochrome P450 is exceeded by that of NADPH-dependent reactions, but usually such reactions need high concentrations of peroxides. In practice, the 'peroxide shunt' utility is limited by fast enzyme inactivation. Since the peroxide shunt reaction of CYPs is independent of the presence of molecular oxygen and, also, of NADPH and redox partners, it may be employed for the development of electrochemical drug screening assay combining CYPs with wide substrate specificity and electrochemical transducers for hydrogen peroxide detection. Metal hexacyanoferrates and especially iron (III) hexacyanoferrate (Prussian blue, PB) are well known for their electro catalytic properties (Karyakin 2001). We employed these electrochemical properties of PB as hydrogen peroxide transducer for the elaboration of amperometric assay based on the peroxide shunt pathway of cytochrome P450 2B4. Due to the high sensitivity of PB-modified electrodes to hydrogen peroxide, it is possible to decrease peroxide concentration for kinetic measurements in the range of 5–50 μM. Addition of substrates of P450 initiated the peroxide consumption, which was detected electrochemically as a decrease of cathodic current of hydrogen peroxide reduction at PB-modified electrode (Fig. 4.3). The proposed electrochemical assay was utilized for the comparative kinetic studies of two drugs (aminopyrine and benzphetamine) as substrates of cytochrome P450 2B4 by the peroxide shunt pathway Shumyantseva et al. 2011b). Peroxide shunt reaction of CYPs, which is a CYP-driven process of organic substrate oxidation through hydrogen peroxide, opens an attractive possibility to employ simple electrochemical techniques for the studies of peroxide-dependent kinetics of these enzymes. Therefore, by combining wide substrate specificity of CYPs and the effective peroxide monitoring system it might be possible to elaborate the novel approach for primary drug screening and/or for the studies of P450 peroxide shunt pathway kinetics with electrochemical techniques.

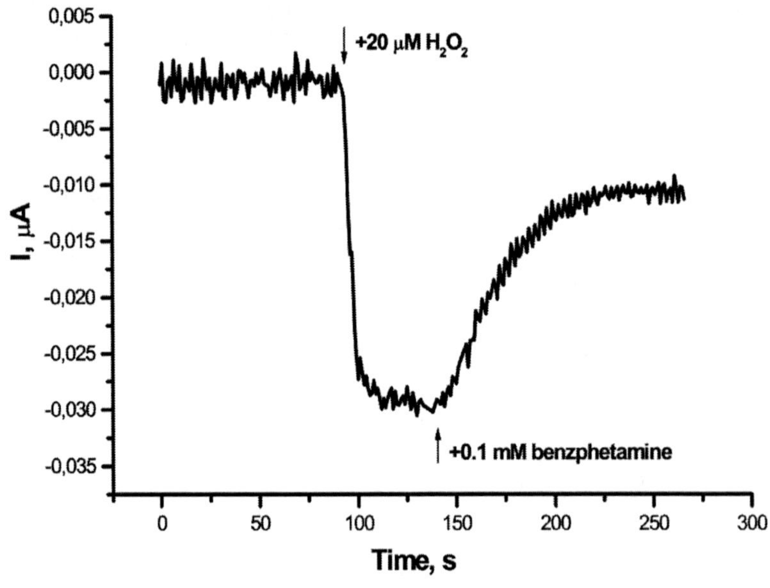

Figure 4.3: Chronoamperometric response of PB/P450 2B4 electrode. Arrows illustrate the points of corresponding substrate addition.

THE INFLUENCE OF VITAMIN B1 ON THE ELECTRO CATALYTIC ACTIVITY OF CYTOCHROME P450 3A4

The problem of drug-drug interaction is intensively studied from different viewpoints. At the same time, the influence of other biologically active compounds on drug-metabolizing enzymes and, also, on pharmacological functions, pharmacokinetics and pharmacodynamics of therapeutic agents has high clinical significance and offers a fertile field for pharmacological research. The enhancement of efficiency and safety of drug therapy is a challenging problem of present-day medicine. This problem may be addressed not only by searching for new therapeutical agents but also thanks to more rational use of already existing medical products. Electrochemical studies were performed to examine the influence of the vitamin B group on the substrate activity of diclofenac (voltaren)—a non-steroidal anti-inflammatory drug used for treatment of arthritis, ankylosing spondylitis and acute muscle pain. Thiamine (water soluble vitamin B1) alone did not demonstrate substrate properties (at 1.5 mM concentration), but blocked the catalytic activity of cytochrome P450 3A4 towards diclofenac (Fig. 4.4). Diclofenac produced 137% increase of catalytic current without thiamine, but in the presence of thiamine only

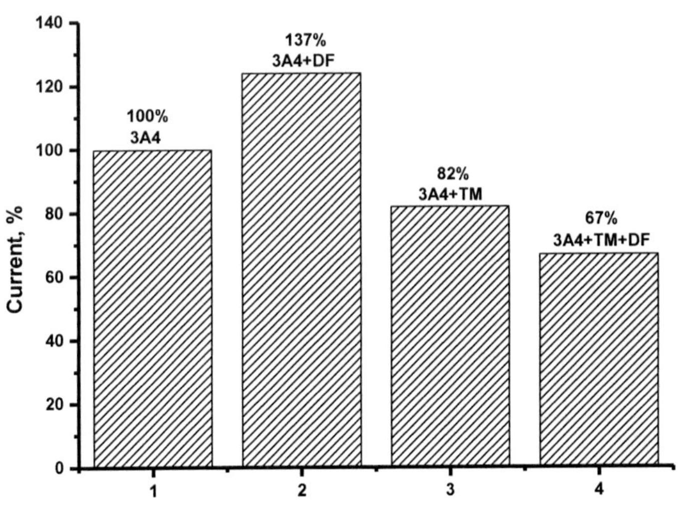

Figure 4.4: Peak intensity (%) of reductive SWV of screen-printed electrodes in aerobic buffer (with baseline correction): DDAB/Au/P450 3A4 (1); DDAB/Au/P450 3A4 +100 µM diclofenac (DF) (2); DDAB/Au/P450 3A4 + 1.5mM thiamine (TM) (3); DDAB/Au/P450 3A4 +1.5mM thiamine (TM), then 100 µM DF (4).

67% of catalytic current can be measured (Fig. 4.4). Since thiamine has nitrogen with an unshared pair of electrons in a pyrimidine heterocycle and sulfur in the thiasole ring it is reasonable to suggest that thiamine interacts with the heme iron of cytochrome P450—similar to the azole inhibitors ketoconazole and itraconazole—while competing with oxygen (Lewis 2001). Oxygen is a cosubstrate of cytochrome P450 enzymes and binds to the heme iron during the catalytic cycle. Thiamine acts as a noncompetitive inhibitor against organic substrate (diclofenac) with the electrochemical inhibition constant Ki = 0.45 ± 0.15 mM. Therefore, thiamine reduces the activity of cytochrome P450 3A4 towards diclofenac. These results suggest that thiamine might influence i.e. slow down, the metabolism of diclofenac in a concentration-dependent manner.

Use of electrochemical methods in combination with nanotechnologies enable in miniaturizing electrodes and measuring devices for the potential application in the form of 'laboratory-on-the-chip' as well as microarrays and point-of-care biosensors (Wang 2006). Further development of methods for the high-throughput search of potential substrates and inhibitors of cytochromes P450 is connected with miniaturization and automation of process, decrease in reagent consumption, assay time, and analysis stages. Finally, this decreases the costs of such studies.

ELECTROCHEMICAL NANOBIOSENSOR FOR EXPRESS DIAGNOSIS OF ACUTE MYOCARDIAL INFARCTION IN UNDILUTED PLASMA

The myocardial infarction biomarker myoglobin was quantified at the biological level in undiluted plasma using developed electrochemical nanosensors with immobilized anti-myoglobin. The method for cardio myoglobin detection is based on direct electron transfer between Fe(III)-heme and electrode surface modified with gold nanoparticles/didodecyldimethylammonium bromide (DDAB/Au) and antibodies. The procedure of myoglobin detection was optimized (pH, incubation times, characteristics of electrodes) to express determination of the marker in serum or plasma. Plasma of healthy donors and patients with acute myocardial infarction was analyzed using electrochemical immunosensors and RAMP® immunoassay. Square wave voltammetry cathodic peak of cardio myoglobin reduction was found to be proportional to myoglobin quantity in plasma as determined by RAMP®.

Cardiovascular disease (CVD) is the cause of nearly half of all the deaths in the western world (Allender et al. 2008) and over half of all the deaths in Russia (http://www.gks.ru). Myocardial infarction (MI)—one of the clinical forms of ischemic heart disease with necrosis of myocardium—leads to a reduction of blood flow. In recent years, electrocardiographic

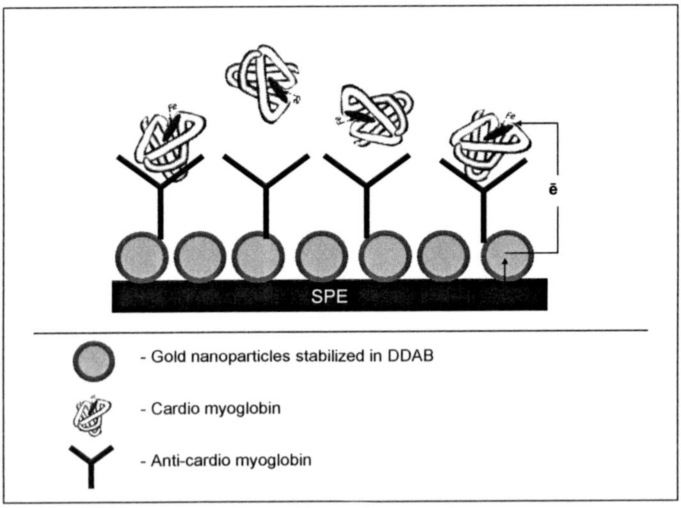

Figure 4.5: Principle circuit of developed cardio myoglobin immunosensor.

investigation has been the main method for establishing MI diagnosis, but only 57% of acute myocardial infarction (AMI) patients reveal electrocardiogram changes. In view of this, the application of cardiac markers (which are always elevated in blood upon the heart occlusion and in CVD) assumes primary importance. The most widespread AMI biomarkers are troponin I, troponin T, creatine kinase-MB and myoglobin (Mb). Cardiac myoglobin was the first nonenzymatic protein used for AMI diagnosis. The small size of the myoglobin molecule (17.8 kDa), facilitating its quick release into circulation (as early as 1–3 hr upon symptom onset), as well as its high sensitivity and high predictive value, makes myoglobin a valuable early screening test for AMI. The initial elevation of cardiac troponin I, troponin T and creatine kinase-MB takes 4–6 hr, with peak release time being registered at 12–24 hr (McDonnell et al. 2009). Sensitive and reliable methods for cardiac markers detection in an early development stage would have a fundamental impact on today's preventive medicine, and myoglobin is one of the very early markers to increase after AMI onset. Myoglobin concentration in blood rises quickly, i.e., 1–3 hr after the appearance of initial symptoms of the disease, and reaches the maximum between 6 and 12 hr. Myoglobin serum concentration returns to the baseline within 24–48 hr. These characteristics of myoglobin help in using them as an early confirmation of AMI (McDonnell et al. 2009). Clinical diagnoses, especially those of heart infarction and heart injury, require reliable test systems. All biochemical methods for AMI diagnosis are based on enzyme-linked immunosorbent assay (ELISA). ELISA has been well known as one of the fastest growing technologies used in clinical diagnostics for detecting specific proteins associated with disease. However, ELISA for cardiac markers needs additional chemicals and cannot be used as point-of-care testing. Portable test-strips and cartridges are able to give qualitative, but not quantitative information on cardiac biomarkers. At present several high-throughputs automated systems are employed in clinics. Commercially available bench-top analyzers and hand-held devices for cardiac biomarkers are Stratus® CS STAT (Dade Behring Inc), i-STAT® (Abbott), Triage® Cardiac Panel (Biosite), RAMP® (Response Biomedical Corp), Cardiac Reader™ (Roche), PATHFAST® (Mitsubishi Chemical Europe GmbH). The current format for these devices is principally two-site immunometric methods, lateral-flow technology and flow-through immunoassay systems. Clinical investigation of heart diseases develops in two main directions: the continuous search for new cardiac biomarkers and implementation of new devises for point-of-care testing with enhanced sensitivity and selectivity (Yang and Zhou 2006).

Plasma proteome is very multiparametric, which makes plasma a very complicated object for electrochemical analysis (Issaq et al. 2007). Cardiac myoglobin cut-off concentrations vary from 70–100 ng/ml (3.93–5.62 nM) to 100–200 ng/ml (5.62–11.24 nM depending on individual specificity of patients (Matveeva et al. 2004). In contrast to widespread application for protein analysis of sandwich immunoassays with secondary (reporter) antibodies labeled with peroxidase, alkaline phosphatase or gold nanoparticles (Piras and Reho 2005), our approach is based on direct detection of specific interaction between anti-myoglobin and hemoprotein and electro catalytic properties of reduced myoglobin towards oxygen (Shumyantseva et al. 2010a, Suprun et al. 2010a). Electro activity of myoglobin on the electrode surface could only be exhibited with modified or specifically structured electrodes. Electrode modification can be accomplished using surfactants (Rusling and Nassar 1993, Guto et al. 2006), polymeric hydrogels (Shen et al. 2002), silk fibroin (Wu et al. 2006), lipids (Nassar et al. 2005), carbon nanotubes (Zhao et al. 2003), metal and metal oxide nanoparticles (Han et al. 2002, Liu et al., 2004, Yang et al. 2006), ionic liquids (Ding et al. 2007), and clays (Wang et al. 2007). Direct electron transfer between electrode modified with gold nanoparticles and didodecyldimethylammonium bromide and myoglobin have been proposed to be utilized for cardiac myoglobin determination. Selectivity of analysis of cardiac human myoglobin is determined by usage of a proper antibody (anti-myoglobin). Analysis takes only 30 min and can be used for express diagnosis of AMI. Quantitative electrochemical determination of myoglobin was performed according to the scheme:

$$Mb\text{–}Fe(III) + e^{-} + H^{+} \leftrightarrow Mb\text{–}Fe(II)$$
$$Mb\text{–}Fe(II) + O_2 \rightarrow [Mb\text{–}Fe(II)O_2] \rightarrow Mb\text{–}Fe(III) + O_2^{-}$$

Analytical characteristics of electrodes were investigated with square wave voltammetry. Upon detection of electroactive biomolecules, gold nanoparticles (AuNP) act as an ensemble of nanoelectrodes on the electrode surface and thus increase the efficacy of electron transfer. Also AuNP act as effective electrocatalysts of Fe(III)/Fe(II) electrode reaction of myoglobin. As was shown earlier, Fe(III)/Fe(II) redox couple of Mb gave standard electron-transfer rate constants that were about 1000-fold higher in liquid crystal films of DDAB on pyrolytic graphite electrodes than in aqueous solutions (Rusling et al. 1993). Gold nanoparticles have a high surface area and good biocompatibility (Wang et al. 2004). Nanostructuring of electrode surface with gold nanoparticles, stabilized by the synthetic membrane-like surfactant DDAB (DDAB/Au), provided effective electron transport between graphite electrode and the Fe(III)/Fe(II) redox couple of myoglobin. The electrochemical method for myoglobin detection, based on electro activity of Fe(III)/Fe(II)-heme and direct detection of specific interaction between anti-myoglobin and heme protein as well as on electro catalytic properties of reduced myoglobin with respect to oxygen, was used for determination of myoglobin in human plasma of healthy donors and patients with AMI. Plasma samples were analyzed using the proposed scheme: SPE/DDAB/Au/anti-HMb + plasma (37°C, 15 min) → 15 min in PBS, pH 6.5 → SWV → SWV analysis of voltammograms. SWV showed no peak for Mb free serum but plasma of patients with acute myocardial infarction (120 ng/ml) had a reductive peak at the position ($E_{SWV} = -250$ mV) that corresponds to the heme protein reduction (Fig. 4.6).

Cathodic peak area and peak height of voltammograms of AMI samples significantly exceeded the corresponding values of SWVs of healthy plasma. Cathodic peak of cardiac Mb reduction was proportional to Mb concentration measured by the cardiac marker point-of–care test RAMP® (Fig. 4.7). Plasma samples with known concentration of myoglobin determined by RAMP were used for checking proportionality of electrochemical response to myoglobin concentration. Electrochemical calibration curve demonstrates this proportionality. Analysis did not require special plasma pretreatments. Small (1 µl) amounts of plasma were used for analysis. Proposed immunosensor has a detection limit of 10 ng/ml (0.56 nM) and a broad range of working concentrations (10 ng/ml–1780 ng/m; 0.56 nM–100 nM) Shumyantseva et al. 2010a, Suprun et al. 2010a).

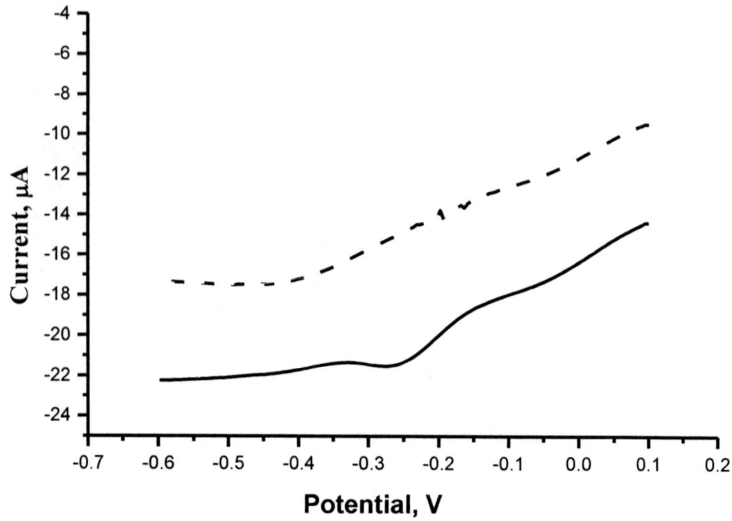

Figure 4.6: Reductive SWV of SPE/DDAB/Au/anti-HMb with 1 μl of serum myoglobin free (– – –) and plasma sample AMI-0001 (120 ng/ml of myoglobin by RAMP). Mb-anti-HMb interaction time 15 min at 37°C. Incubation time in buffer before measurement 15 min. SWV, PBS (pH 6.5), SPE were from *Rusens* with graphite inks from Gwent.

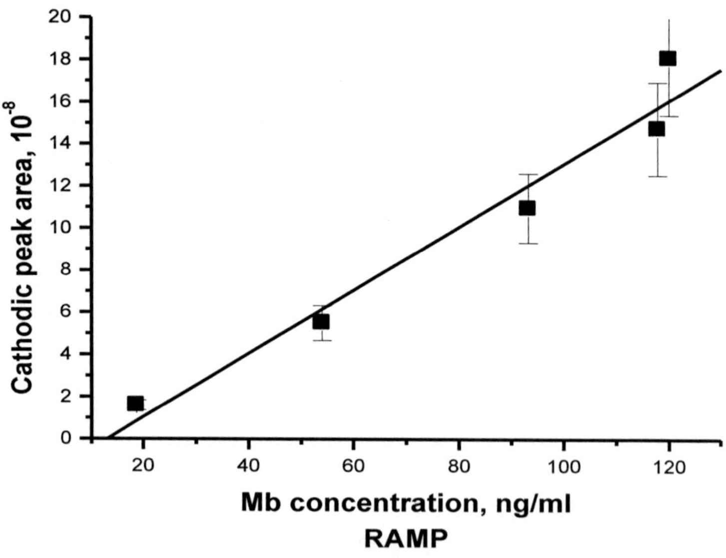

Figure 4.7: Calibration curve of cardio myoglobin in plasma. 1μl of undiluted plasma sample of healthy donors or patients with acute myocardial infarction was dropped onto sensor surface and allowed to dry for 15 min at 37°C. Incubation time in buffer before measurement 15 min. SWV, PBS (pH 6.5), SPE were from *Rusens* with graphite inks from Gwent. Mb concentrations were measured by immunoassay RAMP®. RSD = 0.15 (n=3).

ELECTROCHEMICAL METHODS FOR THE INVESTIGATION OF BIOAFFINITY INTERACTIONS BASED ON GOLD NANOPARTICLES (AuNP) MODIFIED SENSORS

Critical bottleneck of electrochemistry is lacking in electroactivity of a great variety of biologically active compounds or biomarkers in the 'physiological' voltage range. Electrochemical detection and bioaffinity interactions of non electro active biomolecules are based on indirect methods. We propose using AuNP as an analytical tool for the detection of bioaffinity interactions (Fig. 4.8). The oxidation of gold surface (resulted in gold oxide AuO_x formation) upon polarization served as a basis for analytical response. In order to evaluate AuNP surface status differences due to electrode modification, Au/Au–oxide film formation was studied by cyclic (CV) and stripping voltammetry. As a model, the couple thrombin—thrombin binding aptamer was taken. The aptamer was immobilized on a screen-printed electrode modified with gold nanoparticles by avidin-biotin technology. Cathodic peak area was found proportional to thrombin quantity specifically adsorbed onto electrode surface. Calibration curve for thrombin in coordinates lg [thrombin]—peak area of cathodic current (S_{pc}) is sigmoidal as is typical for immunoanalysis, with thrombin detection limit of 10^{-9} M. Linear range corresponds to 10^{-8}–10^{-5} M thrombin concentration or 2×10^{-14}–2×10^{-11} mol/electrode (R = 0.996). This dynamic range corresponds to the physiological level of thrombin in plasma –4.97×10^{-7} M (Zheng et al. 2007). For aptasensor construction we used 10 μM aptamer (2 pmol/electrode), so 10 μM thrombin (2 pmol/

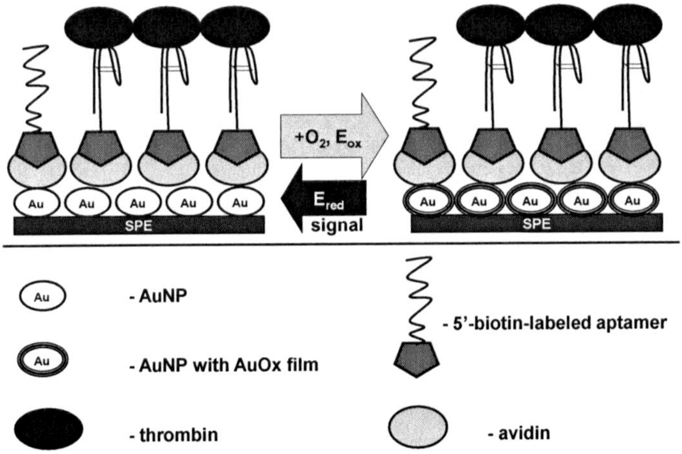

Figure 4.8: Schematic representation of AuNP electrochemical sensing platform for aptamer–thrombin interaction.

electrode) concentrations were expected to be saturated. Specificity of binding was proved with experiments in the presence of trypsin. The difference between AuNP cathodic peak areas in phosphate buffer solution for SPE/AuNP/APT/thrombin and SPE/AuNP/APT/buffer was measured. Specificity of the aptasensor to the target protein was confirmed in experiments with trypsin instead of thrombin binding at the same concentration. Correlation between the cathodic peak area and the logarithm of thrombin concentration was found (Fig. 4.9). In the case of tripsin, the signal has not changed (Suprun et al. 2008).

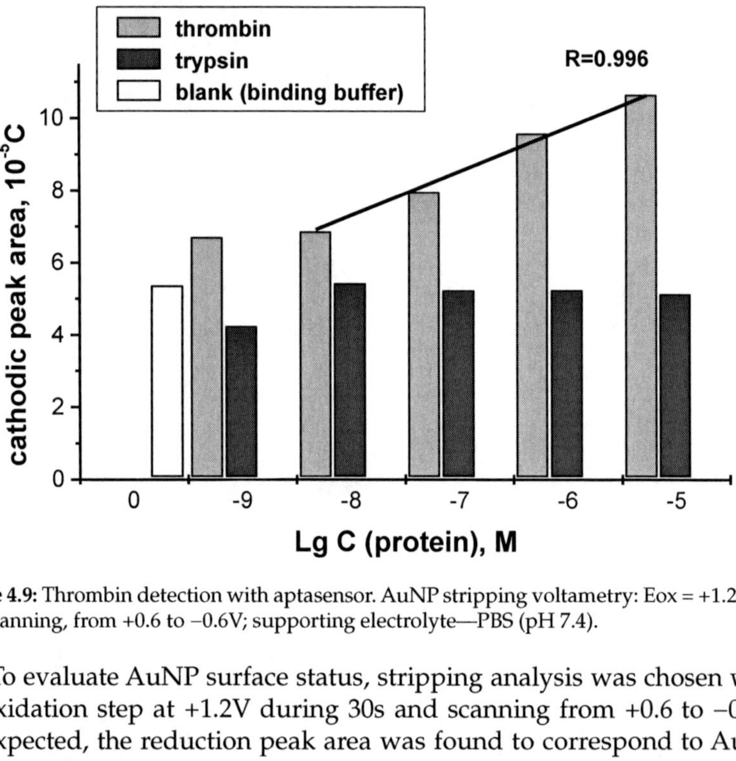

Figure 4.9: Thrombin detection with aptasensor. AuNP stripping voltametry: Eox = +1.2V for 30s, scanning, from +0.6 to −0.6V; supporting electrolyte—PBS (pH 7.4).

To evaluate AuNP surface status, stripping analysis was chosen with preoxidation step at +1.2V during 30s and scanning from +0.6 to −0.6V. As expected, the reduction peak area was found to correspond to AuNP oxidation degree.

Aptamer prefers to adopt G-quarter structure when binding with thrombin (Mao and Gmeiner 2005, Nagatoishi et al. 2007). When the aptamer's conformational changes occur from random coil to the quadruplex, ds-DNA zones arise and facilitate electron transfer from the solution to the electrode surface through the DNA π-stacking interaction (Marques et al. 2003).

The advantages of the proposed approach are: (1) direct measurement of binding events, without amplification stages; (2) small analyzed volume of target protein (2 µl, 2×10^{-14}–2×10^{-11} mol/electrode); (3) wide operating window of thrombin concentrations –4 orders of magnitude –10^{-5}–10^{-9} M; (4) as an electrochemical sensor, the gold nanoparticle platform does not require cumbrous equipment.

Screen printed electrodes modified with silver nanoparticles (AgNP) were also employed as a sensing platform for aptasensor devices. Screen-printed electrodes were modified with silver nanoparticles (AgNP) by electrodeposition technology. The oxidation of silver upon polarization served as a basis for analytical response. DNA-thrombin binding aptamers with various surface concentrations were studied. Aptamers were immobilized onto a screen printed electrode modified with AgNP by using thiol groups at their 5′-termini. Linear range of aptasensor response corresponded to the 10^{-9} M to 10^{-7} M thrombin concentration range and the detection limit was 10^{-9} M (Suprun et al. 2010b). The oxidation of silver upon polarization served as an analytical signal for the sensor. Anodic peak decreased with aptamer immobilization and thrombin binding.

Stripping the voltammetry approach to evaluate nanogold or nanosilver surface status in accordance with sensor modification can be applied to various biomolecules of interest, e.g., to antigen-antibody or other affinity interactions.

ELECTROCHEMICAL METHODS FOR THE INVESTIGATION OF *E. COLI* RESISTANCE TO ANTIBIOTICS

One of the main challenges in medicine and clinical diagnosis, food industry, environmental monitoring is the development of fast, reliable, sensitive and accurate methods for the detection of pathogen microorganisms. Traditional bacterial cell detection methods (culture tests, biochemical tests, microscopy, flow cytometry), are time-consuming, inconvenient and require several handling steps (Alonso-Lomillo et al. 2010, El-Said et al. 2009). Electrochemical methods provide important advantages such as low detection limit, relative simplicity, low cost of equipment, automatic on-line and portable options.

Electrochemical behavior and electrochemical characteristics of *Shawanella oneidensis* cells when immobilized on glassy carbon electrode modified with carbon nanotubes were studied earlier (Peng et al. 2010). The electron transfer between the bacterial cells and the surface of the working electrode took place with the participation of cytochrome c localized on the external membranes of the cells and the oxidizing membrane proteins OmcA and MtrC. The authors registered a sharp cathodic peak on the cyclic voltammetry at –0.42V, with a broad and flat anodic peak at –0.05 V potential on the unmodified glassy carbon electrode. HeLa cells directly immobilized on gold-patterned silicon substrate were used to assess the toxicity of anticancer drugs (hydroxyurea and cyclophosphamide). It was observed that as the concentration of drugs increase, the peak current of cyclic voltammetry response decrease drastically (El-Said et al. 2009).

Our studies aimed at finding the electrochemical activity and electrochemical response of *E.coli* JM109 itself by a direct method when bacterial cells were immobilized onto the electrode surface, modified with DDAB (without mediators). Cyclic voltammetry and square wave voltammetry suggested the current generation could be attributed to *E.coli*'s cell surface redox active protein, such as *b, o, d* cytochromes, flavoproteins, or proteins with FeS-clusters (Figs. 4.10 and 4.11). The emergence of a cathodic peak around −0.4V is the evidence of electron transfer between *E.coli* JM109 outer membrane redox proteins and the surface of the working electrode. The data obtained in our experiments are in agreement with the data of the study *Shawanella oneidensis* cells (Peng et al. 2010) and HeLa cells (El-Said et al. 2009).

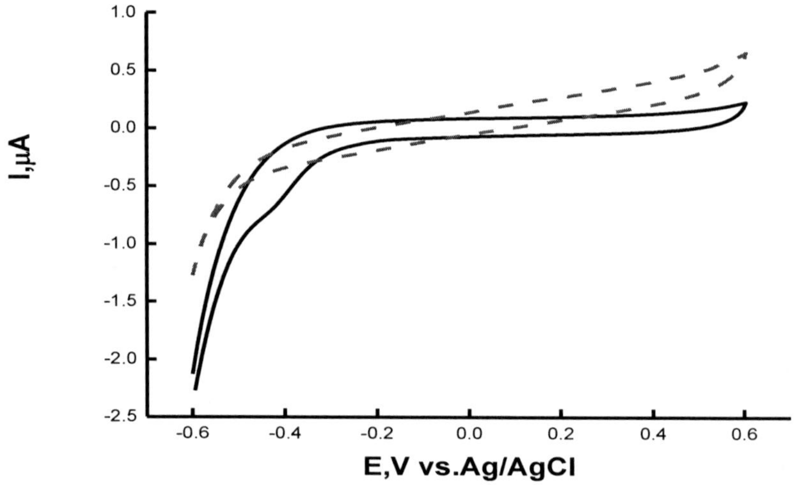

Figure 4.10: Cyclic voltammograms of SPE/DDAB (punctured line) and SPE/DDAB/*E. coli* JM109 (solid line) in aerobic conditions, the scan rate was 100 mV/s. Cell number of $16{\times}10^6$ colony forming units (CFU) per electrode. Electrolyte volume 1 ml, 0.1 M potassium-phosphate buffer, 0.05M NaCl, pH 7.4.

The sensitivity of microbial cells to drugs is potentially useful for the pharmacological industry and clinical medicine. Despite the obvious advantages, the available modern methods of microorganism sensitivity to antibiotic assessment do not make it possible to conduct the microorganism sensitivity to antibiotic research of high quality fast enough owing to some of their disadvantages, the major being the necessity of the pure culture

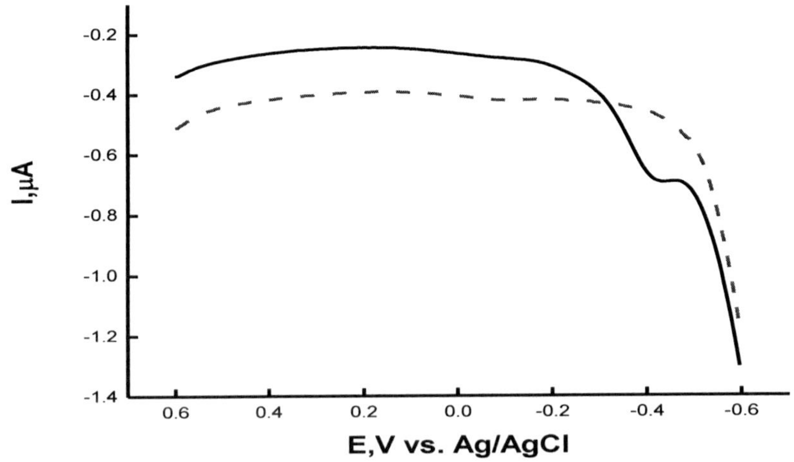

Figure 4.11: Reductive square wave voltammograms of SPE/DDAB electrode (punctured line) and SPE/DDAB /*E.coli* JM109 electrode (solid line) in aerobic conditions (frequency of 10 Hz). Cell number of $16{\times}10^6$ colony forming units (CFU) per electrode. Electrolyte volume 1 ml, 0.1 M potassium-phosphate buffer, 0.05M NaCl, pH 7.4.

extraction. This process takes from 3 to 10 d, which makes it impossible to start and provide the appropriate individual specific treatment at the proper time. Other serious drawbacks of the existing express methods are their high work content, low throughput capacity, inconvenience and/or their high cost (Marks et al. 2007). All of this indicates the necessity of inventing some new techniques to estimate the level of causative microbe sensitivity to antibiotics in a shorter time after the start of the research. Different types of biosensors meet such requirements. Amperometric microbe biosensors have been developed to detect the chemicals suppressing the microorganism growth. Biosensors detecting cyanides and Prussic acid (hydrogen cyanide) by suppressing the respiratory activity of *Nitrosomonas europaea, Thiobacillus ferrooxidans, Saccharomyces cerevisiae,* and *Pseudomonas fluorescents* have been suggested (Marks et al. 2007). The CellSens running system with microbe biosensors has been developed on the basis of *Escherichia coli* cells immobilized on screen printed graphite electrodes, the measuring signal for which is reduction current in the presence of electron transfer mediator—potassium ferricyanide (Farre and Barcelo 2001). However, the above mentioned systems are indirect and require the presence of diffusive electron transfer mediator for assessing the signal, which can result in measurement errors and complicate the experiments.

Based on the electrochemical activity of *E.coli* JM109 bacterial cells a system for determining the sensitivity of microbiological material to antibiotics was proposed. Direct electron transfer between *E.coli* JM109 and screen printed electrode modified with synthetic membrane-like surfactant didodecyldimethylammonium bromide (DDAB), used for the investigation of bacterial cell resistance to antibiotics (Fig. 4.12).

The electrochemical method enables in obtaining data on the antibiotic resistance of the *E.coli* JM109 cells within 2–5 hr. The use of the stencil electrodes can be attributed to the advantages of the system which, in its turn, leads to the miniaturization of the electro analysis, the possibility of the working electrode surface modifying and the wide range of working potentials. It also results in using the minimal volumes of the analyzed material (1–2 µl) and the absence of secondary mediators, which shortens the duration and simplifies the process of analysis. In conclusion, the toxicity of antibiotics with different mechanism of microbial inhibition to *E.coli* JM109 cells has been investigated by using CV and SWV techniques, which allow monitoring electrochemically the viability of living cells.

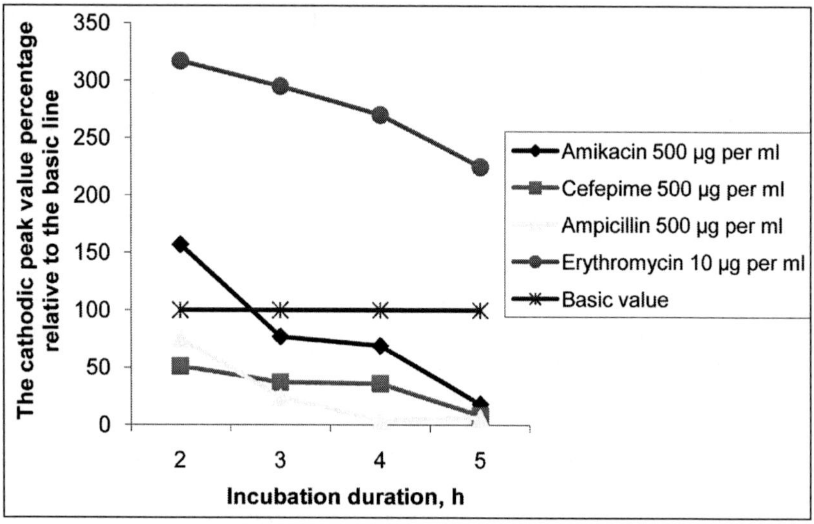

Figure 4.12: Assembled diagram data from experiments carried out on DDAB/*E.coli* JM109 electrodes (1.92×10^6 CFU/electrode) with the addition of antibiotics (Cefepime - Cf, Ampicillin - Amp, Amikacin - Amk, Erythromycin - Em) in volume 5 µl (concentrations of the substances indicated in the title of the axis). As a control 5 µl of buffer was used.

MATERIALS AND METHODS

Electrochemical measurements were carried out using an Autolab PGSTAT12 potentiostat/galvanostat ('Eco Chemie', the Netherlands) with GPES software in a 3-electrode electrochemical cell. All the measurements were taken at room temperature. Electrochemical studies of cytochromes P450 2B4, 1A2 and 3A4 were done in 0.1M potassium-phosphate buffer containing 0.05 M NaCl, pH 7.4; those of CYP51 were done in 0.1 M potassium-phosphate buffer containing 0.04% Triton X-100, 0.05 M NaCl, pH 7.4, and those of CYP11A1, in 0.1 M potassium-phosphate buffer containing 1% Triton X-100, and 0.05 M NaCl, pH 7.4.

Two types of three-pronged screen-printed electrodes (SPE) were used: from *Elcom* (Russia, http://www.elcom-mocsow.ru, with graphite paste from Acheson) and *Rusens* (Russia, http://analyt.chem.msu.ru/, with graphite paste from Gwent). Screen-printed graphite electrodes were used as working electrode and screen-printed Ag/AgCl as reference electrodes (Elkom, Russia). The diameter of working electrode was 2 mm. The potentials are all referred to the Ag/AgCl reference electrode.

Cyclic voltammograms (CV) were registered at the scan rate 10 to 100 mV/s. Parameters used in square wave voltammetry (SWV, reduction, aerobic conditions) were as follows: initial potential, 100 mV; final potential, 600mV: step potential, 5 mV; amplitude, 20 mV; frequency, 10 to 100 Hz. Parameters used in differential pulse voltammetry (DPV) were as follows: the pulse amplitude, 25 mV; initial potential, 100 mV; final potential, 600 mV; step potential, 1 mV; pulse duration, 50 mV.

ACKNOWLEDGMENTS

This work was financially supported by the Federal Agency of Science and Innovations, Ministry of Education and Science of Russian Federation (Contracts no. 02.740.11.0306), RFBR grant no 10-04-91332-DFG_a, and by the Interdepartmental Program "Proteomics in Medicine and Biotechnology".

REFERENCES

Allender, S. and P. Scarborough, V. Peto, M. Rayner, J. Leal, R. Luengo-Fernandez and A. Gray. 2008. European cardiovascular disease statistics. European Heart Network, Brussels.

Alonso-Lomillo, M.A. and J. Gonzalo-Ruiz, O. Domínguez-Renedo, F.J. Muñoz and M.J. Arcos-Martínez. 2008. CYP450 biosensors based on gold chips for antiepileptic drugs determination. Biosens Bioelectron 23: 1733.

Alonso-Lomillo, M.A. and O. Dominguez-Renedo and M.J. Arcos-Martinez. 2010. Screen-printed biosensors in microbiology; a review. Talanta 82: 1629–1636.

Archakov, A.I. and G.I. Bachmanova. 1990. Cytochrome P450 and Active Oxygen. Taylor and Francis, London. UK.

Archakov, A.I. and Y.D. Ivanov, A.V. Lisitsa and V.G. Zgoda. 2009. Biospecific irreversible fishing coupled with atomic force microscopy for detection of extremely low-abundant proteins. Proteomics 9: 1326–1343.

Balakin, K.V. and Y.A. Ivanenkov. 2005. Computational methods for prediction of metabolism of physiologically active substances. Biomed Chim (in Russian) 51: 384–412.

Bellamine, A. and A.T. Mangla, W.D. Nes and M.R. Waterman. 1999. Characterization and catalytic properties of the sterol 14alpha-demethylase from Mycobacterium tuberculosis. Proc Natl Acad Sci 96: 8937–8942.

Bistolas, N. and U. Wollenberger, C. Jung and F.W. Scheller. 2005. Cytochrome P450 biosensors—a review. Biosens Bioelectron 20: 2408–2423.
Bruno, R.D. and C.O. Njar. 2007. Targeting cytochrome P450 enzymes: A new approach in anti-cancer drug development. Bioorg Med Chem 15: 5047–5060.
Carrara, S. and V.V. Shumyantseva, A.I. Archakov and B. Samorì. 2008. Screen-Printed Electrodes based on Carbon Nanotubes and Cytochrome P450scc for Highly-Sensitive Cholesterol Biosensors. Biosens Bioelectron 24: 148–150.
Cirspi, C.L. 1999. Higher-throughput screening with human cytochromes P450. Curr Opin Drug Discov Devel 2: 15–19.
Dalle-Donne, I. and A. Scaloni and D.A. Butterfield. 2006. Redox proteomics. John Wiley & Sons, Inc., Hoboken, New Jersey.
Ding, S-F. and M-Q. Xu, G-Ch. Zhao and X-W. Wei. 2007. Direct electrochemical response of Myoglobin using a room temperature ionic liquid, 1-(2-hydroxyethyl)-3-methyl imidazolium tetrafluoroborate, as supporting electrolyte. Electrochem Commun 9: 216–220.
El-Said, W.A. and C. Yea, H. Kim, B. Oh and J. Choi. 2009. Cell-based chip for the detection of anticancer effect on HeLa cells using cyclic voltammetry. Biosens Bioelectron 24: 1259–1265.
Estabrook, R.W. and K.M. Faulkner, M.S. Shet, and C.W. Fisher. 1996. Application of electrochemistry for P450-catalyzed reactions. Method Enzymol B 272: 44–51.
Farre, M. and D. Barcelo. 2001. Characterization of wastewater toxicity by means of a whole-cell bacterial biosensor, using Pseudomonas putida in conjunction with chemical analysis. Fresenius J Anal Chem 371: 467–473.
Han, X. and W. Cheng, Z. Zhang, S. Dong and E. Wang. 2002. Direct electron transfer between hemoglobin and a glassy carbon electrode facilitated by lipid-protected gold nanoparticles. Biochim Biophys Acta 1556: 273–277.
Harrison, A. and S. Amundson. 2005. Evaluation and management of the acutely dyspneic patient: the role of biomarkers. The American Journal of Emergency Medicine 23: 371–378.
He, X. and L. Zhu. 2006. Direct electrochemistry of hemoglobin in cetylpyridinium bromide film: Redox thermodynamics and electrocatalysis to nitric oxide, Electrochem. Commun 8: 615–620.
Hlavica, P. 2009. Assembly of non-natural electron transfer conduits in the cytochrome P450 system: A critical assessment and update of artificial redox constructs amenable to exploitation in biotechnological areas. Biotechnol Adv 27: 103–121.
Issaq, H. and Z. Xiao and T. Veenstra. 2007. Serum and Plasma Proteome. Chem Rev 107: 3601–3620.
Ivanov, Y.D. and V.M. Govorun, V.A. Bykov and A.I. Archakov. 2006. Nanotechnologies in proteomics. Proteomics 6: 1399–1414.
Joseph, S. and J.F. Rusling, Y.M. Lvov, T. Friedberg and U. Fuhr. 2003. An amperometric biosensor with human CYP3A4 as a novel drug screening tool. Biochem Pharmacol 65: 1817–1826.
Karyakin, A.A. 2001. Prussian Blue and Its Analogues: Electrochemistry and Analytical Applications. Electroanalysis 13: 813.
Kanaeva, I.P. and I.R. Dedinskii, E.D. Scotselyas, A.G. Krainev, I.V. Guleva, I.F. Sevryukova, Ya.M. Koen, G.P. Kuznetsova, G.I Bachmanova and A.I. Archakov. 1992. Comparative study of monomeric reconstituted and membrane

microsomal monooxygenase systems of the rabbit liver: I. Properties of NADPH-cytochrome P450 reductase and cytochrome P450 LM2 (2B4) monomers. Arch Biochem Biophys 298: 395–412.

Lewis, D.F.V. 2001. Guide to Cytochrome P450. Structure and function. Taylor and Francis. London and New York.

Lepesheva, G.I. and N.V. Strushkevich and S.A. Usanov. 1999. Conformational dynamics and molecular interaction reactions of recombinant cytochrome p450scc (CYP11A1) detected by fluorescence energy transfer. Biochim Biophys Acta 1434: 31–43.

Lepesheva, G.I. and M.R. Waterman. 2004. CYP51—the omnipotent P450. Mol and Cellular Endocrinol 215: 165–170.

Lisitsa, A.V. and S.A. Gusev, I.I. Karusina, A.I. Archakov and L. Koymans. 2001. Cytochrome P450 Database. SAR QSAR Environ Res 12: 359–366.

Liu, X. and Y. Huang, W. Zhang, G. Fan, C. Fan and G. Li. 2005. Electrochemical Investigation of Redox Thermodynamics of Immobilized Myoglobin: Ionic and Ligation Effects. Langmuir 21: 375–378.

Liu, S. and L. Peng, X. Yang, Y. Wu and L. He. 2008. Electrochemistry of cytochrome p450 enzyme on nanoparticle-containing membrane-coated electrode and its application for drug sensing. Anal Biochem 375: 209–216.

Liu, S. and Z. Dai, H. Chen and H. Ju. 2004. Immobilization of hemoglobin on zirconium dioxide nanoparticles for preparation of a novel hydrogen peroxide biosensor. Biosens Bioelectron 19: 963–969.

Mao, X. and W. Gmeiner. 2005. NMR study of the folding-unfolding mechanism for the thrombin-binding DNA aptamer d(GGTTGGTGTGGTTGG). Biophys Chem 113: 155–160.

Marks, R.S. and D.C. Cullen, I. Karube, C.R. Lowe and H.H. Weetall. 2007. Handbook of Biosensors and Biochips, John Wiley & Sons, LTD, The Atrium, Southern gate, Chichester, West Sussex, PO19 8SQ, England.

Marques, L.P.J. and I. Cavaco, J.P. Pinheiro, V. Ribeiro and G.N.M. Ferreira. 2003. Electrochemical DNA sensor for detection of single nucleotide polymorphisms. Clin Chem Lab Med 41: 475–481.

Matveeva, E. and Z. Gryczynski, J. Malicka, I. Gryczynski and J.R. Lakowicz. 2004. Metal-enhanced fluorescence immunoassays using total internal reflection and silver island-coated surfaces. Anal Biochem 334: 303–311.

McDonnel, B. and S. Hearty, P. Leonard and R.O'Kennedy. 2009. Cardiac biomarkers and the case for point-of- care testing. Clin Biochem 42: 549–561.

Melanson, S.F. and M.J. Tanasijevic. 2005. Laboratory diagnosis of acute myocardial injury. Cardiovascular Pathology 14: 156–161.

Nagatoishi, S. and Y. Tanaka and K. Tsumoto. 2007. Circular dichroism spectra demonstrate formation of the thrombin-binding DNA aptamer G-quadruplex under stabilizing-cation-deficient conditions. Biochem Biophys Res Commun 352: 812–817.

Nassar, A.E.F. and Y. Narikiyo, T. Sagara, N. Nakashima and J.F. Rusling. 1995. Electrochemical properties of myoglobin embedded in Langmuir–Blodgett and cast films of synthetic lipids. J Chem Soc Faraday Tran 91: 1775–1782.

Nebert, D.W. and D.W. Russell. 2002. Clinical importance of the cytochromes P450. The Lancet 360: 1155–1162.

Ortiz de Montellano, P.R. 2004. Cytochrome P450: Structure, Mechanism, and Biochemistry. 3rd ed. Springer, New York.

Peng, L. and S. You and J. Wang. 2010. Carbon nanotubes as electrode modifier promoting direct electron transfer from Shawanella oneidensis. Biosens Bioelectron 25: 1248–1251.

Persson, K.P. and S. Ekehed, C. Otter, M. Lutz, J. McPheat, C.M. Masimirembwa and T.B. Andersson. 2006. Evaluation of human liver slices and reporter gene assays as systems for predicting the cytochrome P450 induction potential of drugs *in vivo* in humans. Pharm Research 23: 56–69.

Pikuleva, I.A. 2006. Cytochrome P450s and cholesterol homeostasis. Pharmacol Ther 112: 761–773.

Piras, L. and S. Reho. 2005. Colloidal gold based electrochemical immunoassays for the diagnosis of acute myocardial infarction. Sensors and Actuators B: Chemical 111–112: 450–454.

Rudakov, Y.O. and V.V. Shumyantseva, T.V. Bulko, E.V. Suprun, G.P. Kuznetsova, N.F. Samenkova and A.I. Archakov. 2008. Stoichiometry of electrocatalytic cycle of cytochrome P450 2B4. J Inorg Biochem 102: 2020–2025.

Rusling, J. F. and A.-E. F. Nassar. 1993. Enhanced electron transfer for myoglobin in surfactant films on electrodes. J Am Chem Soc 115: 11891–11897.

Sadeghi, S. and A. Fanduzzi and G. Gilardi. 2011. Breakthrough in P450 bioelectrochemistry and future perspectives. Biochem Biophys Acta 1814: 237–248.

Shen, L. and R. Huang and N. Hu. 2002. Myoglobin in polyacrylamide hydrogel films: direct electrochemistry and electrochemical catalysis. Talanta 56: 1131–1139.

Shumyantseva, V.V., T.V. Bulko and A.I. Archakov. 2005. Electrochemical reduction of cytochrome P450 as an approach to the construction of biosensors and bioreactors. J Inorg Biochem 99: 1051–1063.

Shumyantseva, V.V. and T.V. Bulko, G.P. Kuznetsova, A.V. Lisitsa, E.A. Ponomarenko, I.I. Karuzina and A.I. Archakov. 2007a. Electrochemical reduction of Sterol-14 α-demethylase from Mycobacterium tuberculosis (CYP51b1). Biochemistry (Moscow) 72: 658–663.

Shumyantseva, V.V. and T.V. Bulko, Y.O. Rudakov, G.P. Kuznetsova, N.F. Samenkova, F.V. Lisitsa, I.I. Karuzina and A.I. Archakov. 2007b. Electrochemical properties of cytochroms P450 using nanostructured electrodes: Direct electron transfer and electro catalysis. J Inorg Biochem 101: 859–865.

Shumyantseva, V.V. and T.V. Bulko, M.Yu. Vagin, E.V. Suprun and A.I. Archakov. 2010a. Electrochemical Immunoanalysis of Cardiac Myoglobin. Biochemistry (Moscow) Supplement Series B, Biomedical Chemistry 4: 237–242.

Shumyantseva, V.V. and E.V. Suprun, T.V. Bulko, O.V. Dobrinina and A.I. Archakov. 2010b. Sensor systems for medical application based on hemoproteins and nanocomposite materials. Biochemistry (Moscow) Suplement Series B: Biomedical Chemistry 4: 25–36.

Shumyantseva, V.V. and T.V. Bulko, A.Y. Misharin and A.I. Archakov. 2011a. Screening of Potential Substrates or Inhibitors of Cytochrome P450 17A1 (CYP17A1) by Electrochemical Methods. Biochemistry (Moscow) Supplement Series B, Biomedical Chemistry 5: 55–59.

Shumyantseva, V.V. and T.V. Bulko, E.V. Suprun, Y.M. Chalenko, M.Yu. Vagin, Yu.O. Rudakov, M.A. Shatskaya and A.I. Archakov. 2011b. Electrochemical investigations of cytochromes P450. Biochem Biophys. Acta Proteins and Proteomics 1814: 94–101.

Shumyantseva, V.V. and Yu.D. Ivanov, N. Bistolas, F.W. Scheller., A.I. Archakov and U. Wollenberger. 2004 Direct electron transfer of cytochrome P450 2B4 at electrodes modified with non-ionic detergent and colloidal clay nanoparticles Anal Chem 76: 6046–60527.

Smith, D. and N. Sadagopan, M. Zientek, A. Reddy and L. Cohen. 2007. Analytical approaches to determine cytochrome P450 inhibitory potential of new chemical entities in drug discovery. J Chromatogr B 850: 455–463.

Suprun, E. and V. Shumyantseva, T. Bulko, S. Rachmetova, S. Rad'ko, N. Bodoev and A. Archakov. 2008. Au-nanoparticles as an electrochemical sensing platform for aptamer–thrombin interaction. Biosens Bioelectron 24: 831–836.

Suprun, E. and T. Bulko, A. Lisitsa, O. Gnedenko, A. Ivanov, V. Shumyantseva and A. Archakov. 2010a. Electrochemical nanobiosensor for express diagnosis of acute myocardial infarction in undiluted plasma. Biosens Bioelectron 25: 1694–1698.

Suprun, E. and V. Shumyantseva, S. Rakhmetova, S. Voronina, S. Radko, N. Bodoev and A. Archakov. 2010b. Label-free electrochemical thrombin aptasensor based on Ag nanoparticles modified electrode. Electroanalysis 22: 1386–1392.

Tupeinen, M. and U. Jouko, J. Jorma and P. Olavi. 2005. Multiple P450 substrates in a single run: rapid and comprehensive *in vitro* interaction assay. Eur J Pharm Res 24: 123–132.

Udit, A.K. and H.B. Gray. 2005. Electrochemistry of heme-thiolate proteins. Biochem Biophys Res Commun 338: 470–476.

Wang, J. 2006. Electrochemical biosensors: Towards point-of-care cancer diagnostics. Biosens Bioelectron 21: 1887.

Wang, G. and Y. Liu and N. Hu. 2007. Comparative electrochemical study of myoglobin loaded in different types of layer-by-layer assembly films. Electrochim Acta 53: 2071–2079.

Wang, M. and L. Wang, G. Wang, X. Ji, Y. Bai, T. Li, S. Gong and J. Li. 2004. Application of impedance spectroscopy for monitoring colloid Au-enhanced antibody immobilization and antibody -antigen reactions. Biosens Bioelectron 19: 575–582.

Wu, Y. and Q. Shen and S.H. Hu. 2006. Direct electrochemistry and electrocatalysis of heme-proteins in regenerated silk fibroin film. Analytica Chimica Acta 558: 179–186.

Xiao, Y. and C.M. Li. 2008. Nanocomposites: from fabrication to electrochemical bioapplications. Electroanalysis 20: 648–662.

Yang, W. and Y. Li, Y. Bai and CH. Sun. 2006. Hydrogen peroxide biosensor based on myoglobin/colloidal gold nanoparticles immobilized on glassy carbon electrode by a Nafion film. Sensors and Actuators B: Chemical 115: 42–48.

Yang, Z. and D.M. Zhou. 2006. Cardiac markers and their point-of- care testing for diagnosis of acute myocardial infarction. Clin Biochem 39: 771–780.

Zhang, L. and Xu. Liu, C. Wang, Xi. Liu, G. Cheng and Y. Wu. 2010. Expression, purification and direct electrochemistry of cytochrome P450 6A1 from the house fly, *Musca domestica*. Protein Expression and Purification 71: 74–78.

Zhao, G-Ch. and L. Zhang, X-W. Wei and Z-S. Yang. 2003. Myoglobin on multi-walled carbon nanotubes modified electrode: direct electrochemistry and electrocatalysis. Electrochem Commun 5: 825–829.

Zheng, J. and W. Feng, L. Lin, F. Zhang, G. Cheng, P. He and Y.U. Fang. 2007. A new amplification strategy for ultrasensitive electrochemical aptasensor with network-like thiocyanuric acid/gold nanoparticles. Biosens Bioelectron 23: 341–347.

5

Conductive Polymers in Medical Diagnostics

Johannes Daprà,[1] Katrine Kiilerich-Pedersen,[1] Nikolaj Ormstrup Christiansen,[1] Claus Riber Poulsen[1] and Noemi Rozlosnik[1,]*

INTRODUCTION

The demand is growing in the field of medical diagnostics for cost efficient and disposable devices, which demonstrate high sensitivity, reliability and simplicity. Lately, biosensors—in particular conducting polymer-based electrochemical sensors—have demonstrated numerous advantages in areas related to human health, such as diagnosis of infectious disease, genetic mutations, drug discovery, forensics, and food technology, due to their simplicity and high sensitivity.

The major processes involved in any biosensor system are analyte recognition, signal transduction, and readout. Due to their specificity, speed, portability, and low cost, biosensors offer exciting opportunities for numerous decentralized clinical applications—point of care systems.

The ongoing trend in biomedicine is to go smaller. For almost a decade, the buzz word has been *nano*, and the analytical micro devices are now appearing in the clinic. The progress within microfluidic technologies has enabled miniaturization of biomedical systems and biosensors. The down-scaling has several advantages: refined control of fluidics, low sample consumption, applicability to point of care, and low cost.

[1]Department of Micro- and Nanotechnology, Technical University of Denmark, Denmark.
*Corresponding author

Point of care is an emerging field within medical diagnostics and disease monitoring, and eventually disease control. Employing specially designed micro systems, a patient can be monitored continuously at the bed side, and save precious time on commuting between home, doctor and hospital. The technological advancements in the biosensor technology within recent years have accelerated the R&D in point of care devices.

Cost benefit is always an important factor in development of novel medical devices. To reduce the expenses of biosensors, the use of noble metals and cleanroom processing should be kept at a minimum. Therefore, we predict a shift in the usage of gold and platinum to degradable polymer materials. Polymer based microfluidic devices meet the requirements of low cost and mass production, and they are suitable for biosensor applications. One of the most promising conductive polymers is poly(3,4-ethylenedioxythiophene) (PEDOT) and its derivatives due to their attractive properties: high stability, high conductivity (up to 400–600 S/cm) and high transparency.

This chapter will look further into the advantages and applications of all-polymer microfluidic devices for biomedical diagnostics and compare with traditional systems. In many biosensor applications, only one analyte is of interest, and preferentially it should be isolated from an inhomogeneous patient sample. Section 2 provides the reader with an overview of the different novel microfluidic separation techniques in polymeric devices.

Different detection methods are applied in biosensors, some of the promising techniques will be summarized in Section 3. Conductive polymers—primarily PEDOT—are the focus of Section 4. They have many excellent properties and in fact, they can compete with gold in many applications. The focus of Section 5 is sensitivity and specificity of biosensors. High sensitivity and specificity is crucial and can be achieved by functionalization with different molecules. The section will primarily focus on the use of aptamers which is favourable above antibodies. Finally, Section 6 gives an overview of the current status in biosensor development while focusing on ongoing research.

NOVEL MICROFLUIDIC SEPARATION TECHNIQUES FOR SAMPLE PREPARATION

The progress in micro fabrication and lab-on-a-chip technologies is a major field for development of new approaches to bioanalytics and cell biology. Microfluidics has proven successful for cell and particle handling, and the interest in micro devices for separation of particles or cells has increased significantly (Giddings 1993, Nolan & Sklar 1998, Toner & Irimia 2005).

Biological samples comprise a heterogeneous population of cells or particles, which is inconvenient for many biomedical applications, where the objective of study is often just one species. For example, the isolation of CD4+ T-lymphocytes from whole blood is essential to diagnoze human immunodeficiency virus (HIV) (Kuntaegowdanahalli et al. 2009), the isolation of leukocytes is important in drug screening assays, and the isolation of specific micro particles from blood plasma is critical for our understanding of inflammatory diseases. Thus, separation of cells or particles has a wide range of applications within different areas of medicine such as diagnostics, therapeutics, drug discovery, and personalized medicine (Gossett et al. 2010).

Flow cytometry has remained the preferred method for cell sorting by many biologists because the technique is well established and has both high sensitivity and high throughput. Recently, fluorescence based sorting of cells and particles has also been implemented in microfluidic devices.

The microfluidic separation techniques are broadly classified as being either passive or active, depending on the operating principles (Table 5.1). Active separation of particles requires an external force (i.e., electrical power, mechanical pressure or magnetic force), whereas passive separation techniques rely on channel geometry and inherent hydrodynamic forces for functionality (i.e., pillars, pressure field gradient or hydrodynamic force). The following section will introduce a couple of novel separation principles with application in biomedical sensors. For further reading on continuous separation of particles, see review papers by Lenshof and Laurell (2010), Gossett et al. (2010), and Bhagat et al. (2010).

Table 5.1: Active and passive separation technique with application in biomedical sensors.

	Method	**Mechanism**
Active	Acoustophoresis	Acoustic waves
	Optical tweezers	Optical
	Dielectrophoresis	Electric field
Passive	Obstacles	Laminar flow
	Induced lift	Inertial force

Active Separation Techniques

Acoustophoresis

Acoustophoresis is the separation of particles using high intensity sound waves. In a microfluidic system, particles with an induced acoustic standing wave will experience a force towards a node or anti node dependent on their physical properties (Lenshof and Laurell 2010). If two particles suspended in a fluid have opposite acoustic contrast, a separation will occur gathering one at node and the other at anti-node. Generally, rigid particles will have a negative *phase* and move toward the node, whereas air bubbles and lipid vesicles gather at the anti-node (Lenshof and Laurell 2010). After separation, the properties of the laminar flow in the microfluidic channel ensure that particles remain at their position in the channel, hence they can be collected separately with a flow splitter.

Both particles with opposite and similar acoustic contrast can be separated using this technique. The size of particles will influence the time scale. Large particles experience a higher force than smaller ones, and thus gather at the node faster than the small particles. Peterson et al. (2007) described a microfluidic system with three inlets (Fig. 5.1), where a sample composed of different sized particles was introduced at the sides of a microfluidic channel with a sheath fluid in the middle to keep particles in close proximity to channel walls. The system is designed such that an ultrasonic transducer induces a force on the particles, which forces them towards the middle of the channel. Since the larger particles experience a higher force than small particles, the large particles immediately gather at the centre of the channel. Particles are thus allocated proportional to their size. Making use of a flow splitter, particles are separated according to their size. Applying this technique, Peterson et al. (2007) demonstrated separation of a mixture of different sized particles.

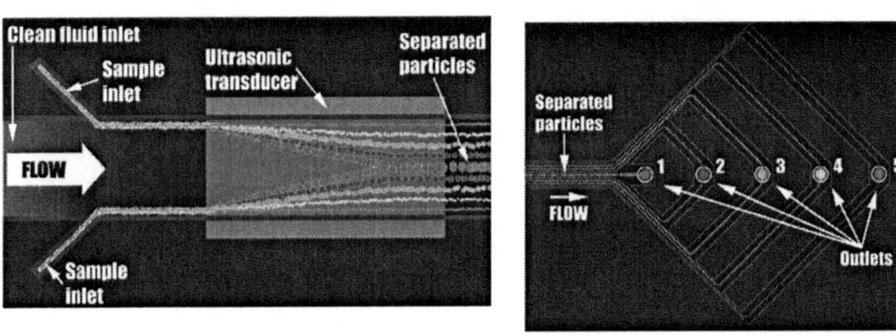

Figure 5.1: Acoustophoresis. (a) Particles entering main channel from two side inlets. Particles are positioned near channel walls because clean sheath fluid is introduced at a third inlet. The flow of particles is controlled by the acoustic waves, which are introduced by an ultrasonic transducer. After this point, the particles distribute proportional to size. (b) Flow splitters are used for separation of different sized particles. Nine fractions of the flow can be gathered at five outlets (Adapted from Peterson et al. 2007).

Color image of this figure appears in the color plate section at the end of the book.

Ion Depletion

Ion depletion is a microfluidic technique for separation and concentration of proteins. As the name indicates, the method is based on ion transfer in a nanofluidic channel (approximately 50 nm in depth). Counter-ions will migrate from the Debye layer through the nanochannel to a higher extent than co-ions, so that a net transfer of counter-ions is transferred from the anodic side to the cathodic side. Thus, the concentration of counter-ions decreases on the anodic side and an increase is achieved on the cathodic side. If a protein in solution is part of the co-ion population, this protein will be trapped in a plug on either side of the ion depletion region, and is hence separated from the bulk solution. The principle of ion depletion is illustrated on Figs. 5.2 and 5.3 (Wang et al. 2005).

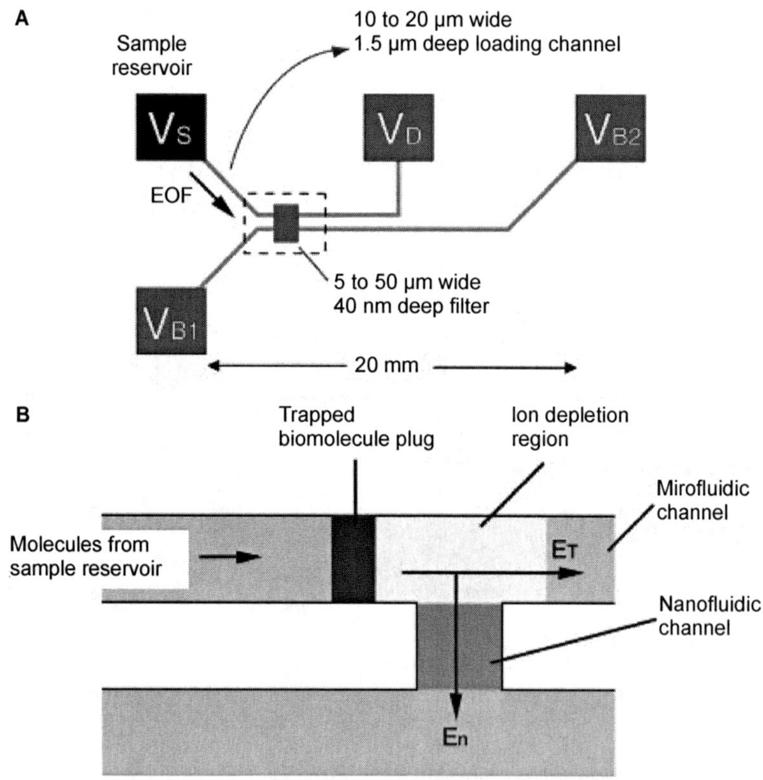

Figure 5.2: Nanofluidic protein concentrating device by ion depletion: (A) Layout of the device. (B) Schematic diagram showing the concentration mechanism. Once proper voltages are applied, the trapping region and depletion region will be formed as indicated. The E_T specifies the electrical field applied across the ion depletion region, while the E_n specifies the cross nanofilter electrical field (Adapted from Wang et al. 2005).

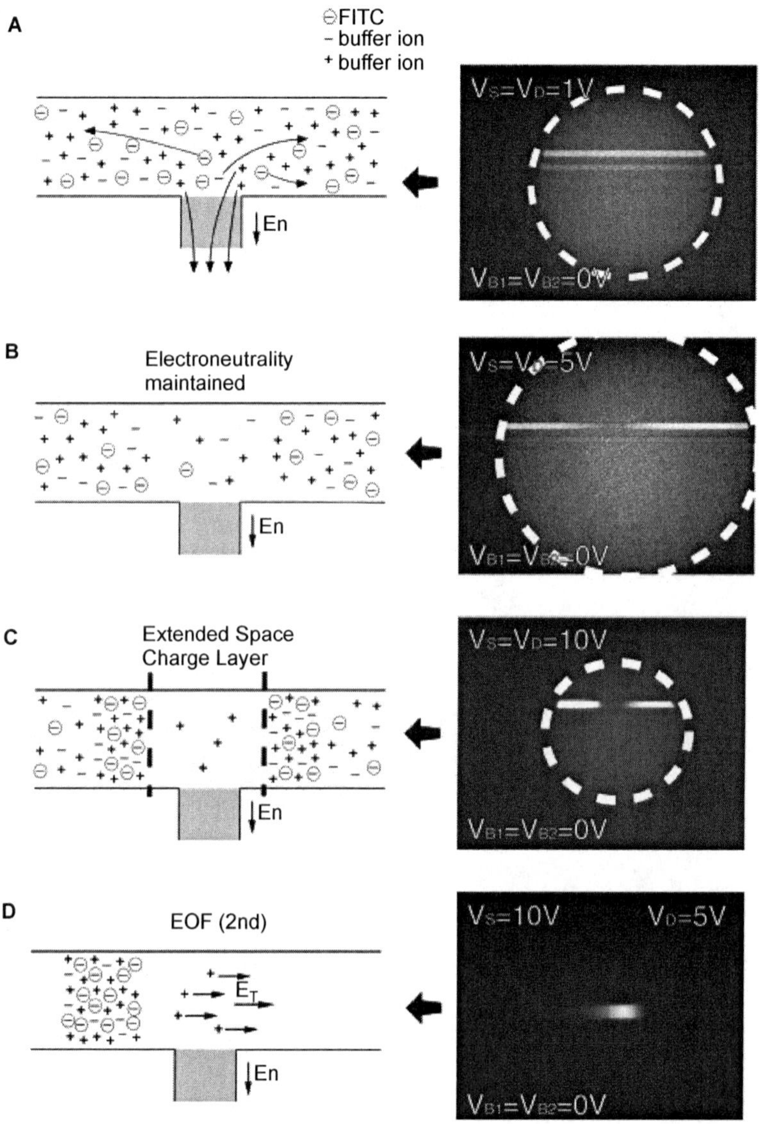

Figure 5.3: Mechanism of preconcentration in the nanofilter device (A) No concentration polarization is observed when a small electrical field (E_n) is applied across the nanofilter. (B) As the E_n increases, the transport of ions becomes diffusion-limited and generates the ion depletion zone. However, the region maintains its electroneutrality. (C) Once a strong field (E_n) is applied, the nanochannel will develop an induced space charge layer, where electroneutrality is no longer maintained. (D) By applying an additional field (E_T) along the microfluidic channel in the anodic side (from V_S to V_D), a nonlinear electrokinetic flow (called electroosmosis of the second kind) is induced, which results in fast accumulation of biomolecules in front of the induced space charge layer (Adapted from Wang et al. 2005).

Passive Separation Techniques

Obstacles

Obstacles arranged in microfluidic channels are commonly applied for preventing particles from entering certain areas or used to manipulate the flow of fluid in a microchannel. Deterministic lateral displacement is a method for size separation of particles or cells, accomplished by placing posts asymmetrically in a microchannel (Fig. 5.4) and thus forcing particles of different sizes to follow different flow paths.

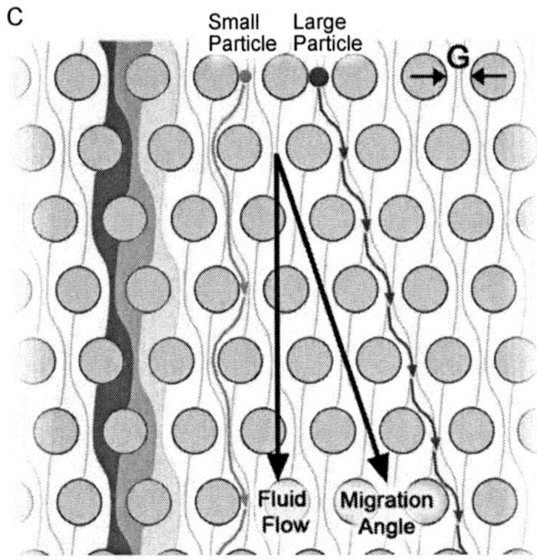

Figure 5.4: Deterministic lateral displacement (Adapted from Gossett et al. 2010).

Spiral Microchannels

Separation of particles in a spiral microchannel was described by Kuntaegowdanahalli and colleagues (2009) (see Fig. 5.5).

It is a passive separation technique based on the centrifugal force. Centrifugal based techniques have been demonstrated using flows in curvilinear microchannels (Gregoratto et al. 2007, Seo et al. 2007). In general, the flow of fluid through a curvilinear channel experiences a centrifugal acceleration, directed radially outward. The channel geometry gives rise to vortices, which are exploited for separation of different sized particles. Particles in the centre of the channel will experience a drag away from the centre, whereas particles in the proximity of the channel walls experience repulsion from the walls. Consequently, particles align at four equilibrium positions in the channel and different sized particles can thus be collected at different outlets (Bhagat et al. 2008, Di Carlo et al. 2007).

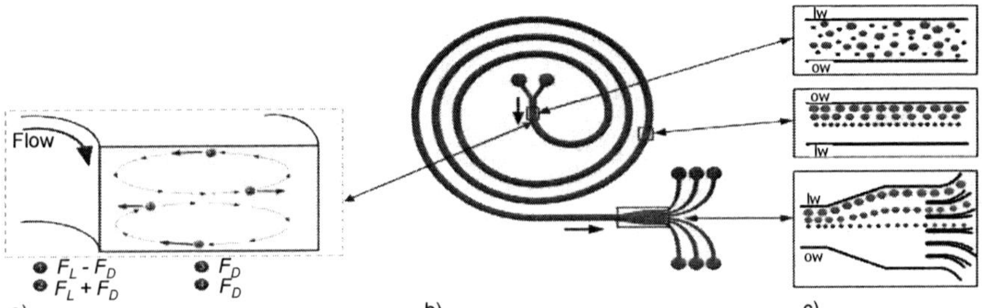

Figure 5.5: Spiral microchannel. (a) Neutral buoyant particles suspended in a medium in a spiral shaped channel experience forces and drag. Resultantly, particles redistribute within the microchannel. (b) Schematic representation of spiral channel for particle separation. (c) Different sized particles equilibrate at different positions in microchannel, and are collected at different outlets (Adapted from Kuntaegowdanahalli et al. 2009 and Bhagat et al. 2008).

Color image of this figure appears in the color plate section at the end of the book.

ELECTRICAL DETECTION METHODS

Modern biosensors for medical diagnostics must be specific, quick, and reproducible at reasonable cost. Electrical sensing is one of the simplest detection methods in microsystems. The electrical detection has traditionally received the major share of the attention in biosensor development. Such devices produce a simple, inexpensive and yet accurate and sensitive platform for patient diagnosis.

The name 'electrochemical biosensor' is applied to a molecular sensing device which intimately couples a biological recognition element to an electrode transducer. The purpose of the transducer is to convert the biological recognition event into a useful electrical signal at the electrode-solution interfaces.

Microelectrodes are powerful and versatile tools in the study of electrochemical processes of mechanistic and/or analytical interest. They experience high mass transport rates but little interference from interfacial capacitance or solution resistance effects. These advantageous properties are due to the small size of these devices. Microelectrodes can work with very small sample volumes enabling the detection of very small amounts of material. The improved mass transport properties facilitate the measurement of higher exchange current densities and electron transfer rate constants and also allow the study of fast coupled chemical reactions.

Electrochemical systems are extremely sensitive to the processes that take place on the surfaces of the electrodes, and in this sense the electrodes are direct transducers in biomedical applications. Several types of electrochemical methods are used in biosensors; the two most common ones are the amperometry and impedance spectroscopy (EIS) (Lazcka et al. 2007). Recently, all-polymer field effect transistors for biosensing have been introduced (Lee et al. 2010).

Amperometry

Amperometry is a method of electrochemical analysis in which the signal of interest is a current that is linearly dependent upon the concentration of the analyte. As certain chemical species are oxidized or reduced (redox reactions) at the electrodes, electrons are transferred from the analyte to the working electrode or to the analyte from the electrode. The direction of flow of electrons depends upon the properties of the analyte and can be controlled by the electric potential applied to the working electrode.

Amperometric biosensors operate by applying a constant potential and monitoring the current associated with the reduction or oxidation of an electroactive species involved in the recognition process. The amperometric biosensor is attractive because of its high sensitivity and wide linear range.

Conductivity and Impedance Spectroscopy

Electrochemical impedance spectroscopy (EIS) combines analyses of both the resistive and capacitive properties of materials, based on the perturbation of the system by a small-amplitude sinusoidal AC signal. The impedance of the system can be scanned over a wide range of AC signal frequencies. The amplitude of the current, potential signals and the resulting phase difference between voltage and current dictates the system impedance. Therefore, the impedance signal is dependent on the nature of the system under study.

Equivalent circuit models fitted to the impedance curves are useful tools for characterizing the system. Although this methodology is widely accepted because of ease of use, extreme care must be taken to ensure that the equivalent circuit obtained makes physical sense. An advantage of EIS compared to amperometry is that redox labels are no longer necessary, which simplifies the sensor preparation.

Organic Field Effect Transistors

Organic field effect transistors (Organic FETs) have a potential of being the active matrix for many electronic devices, including biosensors for biological material. An organic field-effect transistor consists of a source and drain electrode, an organic semiconductor (which is in this case a conductive polymer), a gate dielectric, and a gate electrode. A number of different studies have demonstrated conductance-based sensors employing a molecular receptor layer immobilized on the surface of a semiconductor device. The receptor molecules provide the means to achieve highly selective sensing because they can be engineered to have much higher binding affinities with the desired target molecules than the other species in the analyte solution (see Section 4). Although the organic FET is a promising candidate for biosensor applications, optimization of the device structure and operating conditions is still required.

CONDUCTIVE POLYMERS FOR SENSING

A major cost factor in electrochemical biosensors is the electrode material—often noble metal—demanding extensive production steps in cleanroom facilities. To cut down on these expenses there is a trend to utilize conductive polymers for sensing. This section will give an introduction to advantages of conductive polymers compared to noble metals, and guide through the considerations associated with selecting an appropriate polymer material for biosensor applications.

Polymers or Metals?

The application of polymers as supporting materials in microfluidic systems is well established; however the electronic sensing units in most chips are fabricated from metallic conductors such as platinum or gold.

Biocompatibility, high sensitivity and specificity are a demand in modern medical biosensors. Biocompatibility is required because some biological applications involve living cells, bacteria or virus. High specificity and sensitivity is essential for detecting highly diluted analytes in biological samples, because the samples contain a cocktail of similar components, which can influence a measurement. All of these requirements can be fulfilled by the metal electrode materials such as solid platinum or gold (Prodromidis and Karayannis 2002). Though, a major disadvantage of the noble metals is the high cost, which is continuously increasing.

Conjugated polymers are an alternative to the traditional electrode materials. The electronic structure of these compounds gives them properties similar to inorganic semiconductors. In 1977, Shirakawa et al. discovered that doping polyacetylene with halogens increased the conductivity by up to four orders of magnitude. The following research on this topic by Shirakawa, MacDiarmid and Heeger was awarded with the Nobel Prize in chemistry in 2000.

Over the years, electronically conductive polymers have been proposed for many applications (Jagur-Grodzinski 2002, Olson et al. 2010)—from biomedical sensors to nanowire integration in photovoltaic cells or printable RFID antennae—yet only few have made it to the market. Among those are electrochromic coatings for windows, antistatic coatings, organic light emitting diodes (OLEDs), and corrosion protection for metals or surface finish for printed circuit boards (Groenendaal et al. 2000, Gustafsson et al. 1994, Wessling 2001).

The immediate advantage of conductive polymer electrodes is the much lower cost of raw materials and the inexpensive production steps. Certain polymers offer high biocompatibility and options for modifying the properties by varying side groups. This can be useful for probe immobilization, which is a crucial procedure in biosensors. Conductive polymers allow a broad range of chemical modifications for covalent attachment of enzymes, antibodies, DNA or other bioprobes (Sarma et al. 2009, Teles and Fonseca 2008).

In summary, replacing metals with polymers as electrode material does not only limit the cost on the materials themselves, but also allows for the inexpensive mass production by modern ink-jet printing methods (Loffredo et al. 2009, Mabrook et al. 2006) or agarose stamping (Hansen et al. 2007).

Polymer Selection

As mentioned in Section 3.1, biocompatibility is a very important factor in selecting an appropriate polymer. Biocompatibility is mainly influenced by the intrinsic toxicity of a material but also by hydrophilicity. Many conjugated polymers suffer from degradation because of irreversible oxidation processes, or they lose their conductive properties over time. A constant and reliable signal is crucial for sensor devices, and accordingly the polymer should be stable over a certain period of time.

In order to provide a good signal to noise ratio in electrochemical measurements, a low ohmic resistance (i.e., high conductivity) is preferred. Currently, these requirements are met by few polymers on the market.

Polypyrrole

The physical properties of polypyrrole (PPy, Fig. 5.6(a)) make it suitable for biosensor applications. PPy has high decomposition temperature (180–237°C), glass transition temperature (Tg, 160–170°C), and relatively high conductivity of up to 3 $S{\cdot}cm^{-1}$ (Biswas and Roy 1994). Besides, PPy has a good environmental stability and different facile processing methods (Wang et al. 2001).

In 2005, Dubois et al. developed a PPy based biosensor for label-free detection of peanut agglutinin. The lactosyl probe unit was immobilized on a biotinylated PPy film via avidin bridges. Their findings demonstrated that the bioprobe could be immobilized directly on the functionalized electrode surface, facilitating label-free detection by electrochemical methods.

There are different strategies to functionalize the electrode surface, and another approach was described by Campbell et al. (1999). They incorporated human erythrocytes into the PPy matrix, and upon capture and binding of Anti-Rhesus (D) antibody, a resistance change could be detected. Other techniques will be discussed in Section 4.

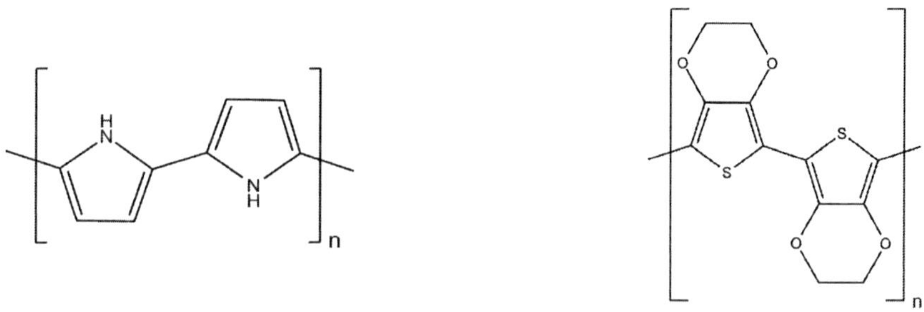

Figure 5.6: Monomer units of (a) polypyrrole (PPy) and (b) poly(3,4-ethylenedioxythiophene) (PEDOT).

Poly(3,4-ethylenedioxythiophene)

Improved properties compared to PPy were found for poly(3,4-ethylenedioxythiophene) or PEDOT. It is either chemically or electrically polymerized from the commercially available monomer 3,4-ethylenedioxythiophene. As can be seen in Fig. 5.6(b), it has some structural similarities with PPy. PEDOT has exceptional high conductivity (up to 600 $S{\cdot}cm^{-1}$), high environmental stability and is biocompatible and transparent for visible light. The most common dopants used for PEDOT are poly(styrene sulphonate) (PSS) and tosylate. Several new methods have been also used to enhance the sensitivity, applicability and/or specificity of these sensors based on PEDOT, for example the incorporation of nanoparticles into the polymer matrix, ink jet printing/patterning of the conducting polymers, molecular imprinting for specific detection, the creation of organic transistors from conducting polymers to improve sensor sensitivity, or the embedding of cells into the conducting polymer matrix for direct stimulation (Rozlosnik 2009).

The works by Balamurugan and Chen (2007) and Vasantha and Chen (2006) show the high potential and superior qualities of PEDOT, and this conductive polymer has been employed in a number of biosensor micro devices.

An interesting study was presented by Kumar et al. (2006). A biosensor was developed to determine the concentration of the important mammalian neurotransmitter, dopamine via an electrochemical process. Since the concentration of ascorbic acid is around a thousand times higher than dopamine in a biological sample, and the two analytes have similar electrochemical potentials, the challenge was to measure the concentration of dopamine in the presence of ascorbic acid. Kumar et al. (2006) employed glassy carbon electrodes coated with PEDOT, and their findings demonstrated significant peak separation and improved anti-fouling properties compared to the more common electrode material glassy carbon, making PEDOT a good candidate for further applications in this field.

Glucose detection for blood sugar monitoring of diabetes patients is a huge and growing market for disposable biosensors. The established commercial systems make use of metal electrodes (typically Pt) coated with a gel containing the enzyme glucose oxidase, and the effectively measured agent is thus the oxidation product, hydrogen peroxide (H_2O_2). In contrast to the direct oxidation of dopamine on the electrodes in the example above, this indirect detection of glucose is more complicated. Considering the current market price of platinum of about 41 €/g (http://platinumprice.org), replacing the electrode material with a low cost polymer such as PEDOT seems sensible. Park et al. (2008) imprisoned glucose oxidase in hollow PEDOT micro-tubules on an indium-tin-oxide (ITO) glass surface (Fig. 5.7). In this configuration, the enzymes are surrounded by the electrode, and therefore their activity is not constrained by immobilization on a surface or incorporation into a polymer. Although the performance of this biosensor cannot meet the requirements of a classic system, it can be refined by increasing the enzyme density or improving the conductivity.

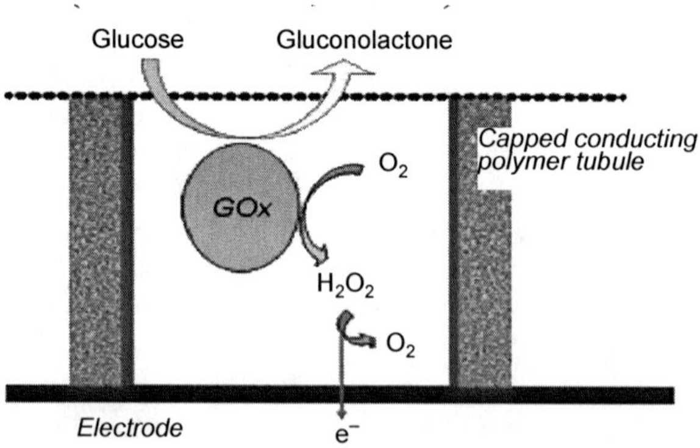

Figure 5.7: Glucose oxidase is imprisoned inside a PEDOT microtube covered with a non-conductive polymer (Park et al. 2008).

Many biosensors for pathogen detection are based on antibodies as probes, and deliver an indirect signal. These immunosensors require a fluorescently tagged second antibody, which reacts with occupied immobilized antibodies in a so-called sandwich assay.

A different approach was tested by Kim et al. (2010), who worked on the development of a point of care system for prostate specific antigen/ α1-antichymotropsin (PSA-ACT) complex detection. This cancer marker is associated with prostate tumours and important for preoperative diagnosis and screening. Instead of using the conventional optical methods, they constructed an organic electrochemical transistor (OECT) based on PEDOT. The antigen was captured by immobilized antibodies on the conductive polymer. For signal enhancement, a secondary antibody with a covalently tethered gold nanoparticle was used. The system provided a detection limit as low as 1 $pg{\cdot}mL^{-1}$ and is thus sensitive enough for reliable PSA-ACT analysis.

PEDOT Derivatives

A field effect transistor (FET) based biosensor was demonstrated by Xie et al. (2009). The working principle is fundamentally different, considering it uses conductive polymer nanowires, which were electropolymerized between two gold electrodes. For minimizing the distance between polymer and binding event it was necessary to couple the probe (an aptamer, see also Section 4.3) directly to the electrode material. Normal PEDOT offers no possibility for covalent bonding of other molecules, so a derivative bearing a carboxylic acid group was used. With this functional group the oligonucleotide for thrombin detection was attached with a simple 1-ethyl-3-(3-dimethylaminopropyl)car-bodiimide/*N*-hydroxysuccinimide procedure (see Section 4.3, (EDC/NHS)). Thrombin binds specifically to the aptamers and becomes immobilized on the surface. The positively charged protein influences the transistor, so that the current flow changes. This type of biosensor has a broad dynamic range covering the physiologically interesting thrombin concentration range from a few to several hundred nanomoles.

Other PEDOT derivatives have also been investigated (Akoudad and Roncali 2000, Ali et al. 2007, Daugaard et al. 2008). The structural formulas of the most commonly used monomers are shown in Fig. 5.8; PEDOT-OH is more hydrophilic than normal PEDOT, and the azide modified PEDOT-N3 polymerizes slowly and has decreased conductivity. The only commercially available monomer is (2,3-dihydrothieno[3,4-*b*][1,4]dioxin-2-yl)methanol (commonly known as hydroxymethyl-EDOT or EDOT-OH) (8(a)), and it can be used as a basis for further modifications.

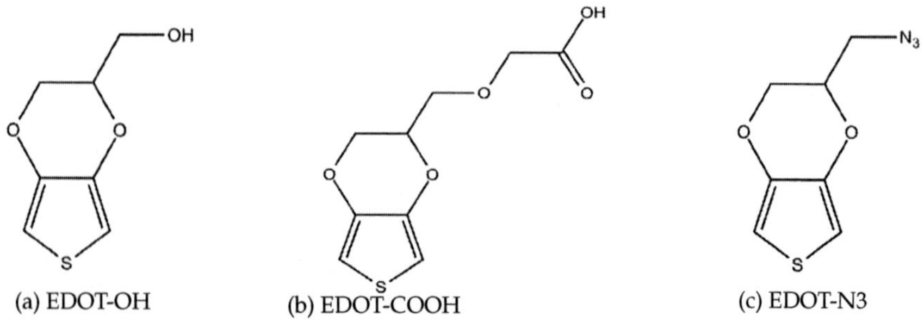

(a) EDOT-OH (b) EDOT-COOH (c) EDOT-N3

Figure 5.8: Different derivatives of 3,4-ethylenedioxythiophene (Ali et al. 2007, Daugaard et al. 2008).

Immobilization Methods

There are different techniques for immobilization of biomolecules (e.g., DNA) on an electrode surface. The most popular methods are formation of a biotin-streptavidin complex, formation of different covalent bonds like esters or amides, or click chemistry.

Biotin-Streptavidin Complex

Streptavidin is a protein consisting of four identical subunits, each of which has an extremely high affinity for biotin. A biotinylated surface can be coated with streptavidin so it offers reactive sites for fixation of likewise biotin tagged (bio)molecules. The biotin-streptavidin interaction is one of the strongest non-covalent bonds in nature and it is very specific. Moreover, the system is easy to handle and very biocompatible.

Despite the many advantages of streptavidin, a major drawback is the instability at low or high pH values, and high temperature. For some detection methods the rather thick protein layer between electrode and probe can substantially decrease the sensitivity of the sensor.

Covalent Bonding

Different activation methods have been used for a long time in chemistry, which require the availability of certain functional groups on the surface. The activation of a carboxylic acid group with 1-ethyl-3-(3-dimethylaminopropyl) carbodiimide (EDC) and *N*-hydroxysuccinimide (NHS) is often applied for amide-bond formation under mild conditions and can be used for binding molecules bearing free amino groups (Balamurugan et al. 2008, Xie et al. 2009).

For hydroxyl functionalized polymers and target molecules, a technique from DNA synthesis can be employed. The alcohol groups are activated with phosphoramidites to form a phosphoester, which then reacts with another hydroxyl moiety and links the target molecules covalently to the surface (Pirrung 2002).

Click Chemistry

A very elegant approach for probe immobilization is the usage of so called 'click-chemistry'. In the Cu-catalysed Huisgen-type 1,3-cycloaddition suggested by Daugaard et al. (2008) an azide reacts in high yield with an alkyne to form a five-membered heterocycle. This bond is very stable and also the precursors have advantages such as stability toward hydrolysis and dimerization or ease of introduction (Kolb et al. 2001).

However, the azide functionalization of PEDOT downgraded its conductive properties significantly and the remaining Cu catalyst could influence biological systems.

ELECTRODE FUNCTIONALIZATION

Functionalization of electrodes is essential for achieving high sensitivity and specificity of electrochemical biosensors. This section provides an overview of the current trend in electrochemical sensors for medical diagnostics.

Recognition of Pathogens

Point of care diagnostic devices present a viable option for rapid and sensitive detection and analysis of pathogens. Biosensors can play an important role in the early diagnosis of acute viral disease and confine the spread of virulent disease outbreaks. Biosensors can also play an important role in early detection and diagnosis of cancer and autoimmune disorders based on specific biochemical markers.

As discussed in Section 2, separation and isolation of large quantities of a specific analyte would be preferable for many medical applications.

Patient samples comprise of a heterogeneous population of particles and cells, hence challenging the isolation of a single species in a high background concentration. For this reason, biosensors must be very specific and sensitive, allowing precise detection of very small quantities.

Antibodies

Many techniques for preparing functional biological surfaces for studies of cells, viruses or disease markers have been described in the literature. Refer to Section 4.3 for an overview of different coupling methods.

Immunoglobulins (IgG) are large Y-shaped proteins produced by the immune system, and are most abundant in blood plasma. Two identical antigen binding sites are formed from several loops of the polypeptide chain. These loops allow many chemical groups to close in on a ligand and link to it with many weak (non-covalent) reversible bonds. An antibody-antigen bond is highly specific because of the molecular structure of the protein.

Antibodies are the most common recognition molecules in biosensors. It is a naturally occurring protein and can only be produced in a host against immunogenic substances, giving rise to batch variation and a limited target range. For research purposes, monoclonal or polyclonal antibodies can be applied as a recognition molecule. Typically, monoclonal antibodies will ensure a higher specificity than polyclonal antibodies.

In medical sensor applications, functional orientation of the antibodies on the surface is crucial to ensure high sensitivity and specificity. It can be achieved by immobilizing the proteins on a supporting layer of protein A.

Aptamers

For many years, antibodies have been applied for surface functionalization in biosensors, ensuring specificity and sensitivity of sensors. Artificial nucleic acid ligands—known as aptamers—can cover the same field of application as antibodies. In the recent years, the use of aptamers has increased (Han et al. 2010, Syed and Pervaiz 2010), and they are in many ways superior to antibodies, as will be discussed in this section.

The Properties of Aptamers

Aptamers are oligonucleotides with a typical length of 40 to 80 base pairs, and were discovered in the 1980's as naturally occurring regulation elements in prokaryotic cells. They showed high affinity for viral and cellular proteins.

In 1990, Tuerk and Gold developed a convenient process for *in vitro* aptamer production, the so-called systematic evolution of ligands by exponential enrichment (SELEX, see in Section 5.3.2).

Aptamers are in many ways better than antibodies as is summarized in Table 5.2. The affinity for the target molecules of aptamers is similar to antibodies, and in some cases even higher compared to antibodies. The specificity is also higher for aptamers, as they can distinguish between targets of the same family, like it was shown for the molecules caffeine and theophylline (Zimmermann et al. 2000). Selection and production of the nucleic acid ligands can be done *in vitro*, and once the correct sequence has been determined, the oligonucleotide can be synthesized in an automated

Table 5.2: Differences between aptamers and antibodies. Advantages are emphasized (Lee et al. 2008).

	Aptamers	**Antibodies**
Affinity	Low nM—pM	Low nM – pM
Specificity	High	High
Production	In vitro chemical process	In vivo biological process
Target range	Wide: ions—whole cells	Narrow: immunogenic compounds
Batch to batch variation	Little or no	Significant
Chemical modification	Easy and straightforward	Limited
Thermal denaturation	Reversible	Irreversible
Shelf life	Unlimited	Limited

chemical procedure. The range for possible target molecules is very wide and—in comparison with the mentioned biomolecules—comprises all kinds of smaller ions, organic compounds and even whole cells. Contrary to antibodies, aptamers can be selected against toxic compounds.

Due to the chemical synthesis, there is no significant batch variation and it allows for easy chemical modification, like attachment of certain end groups for surface immobilization.

Reversible thermal denaturation makes aptamers potentially recyclable and their very high stability promises a long shelf life (Lee et al. 2008).

The SELEX Process

Aptamer production is accomplished in the SELEX (**S**ystematic **E**volution of **L**igands by **Ex**ponential Enrichment) process (see Fig. 5.9). A pool of single stranded oligonucleotides with a random section of about 25 to 70 base pairs (the library) is incubated with the target molecule. Some nucleic acid strands will interact with the target molecule and form strong non-covalent bonds. Target-DNA-complexes are partitioned from unbound DNA. After dissolution of the complex, the selected oligonucleotides are amplified in a standard PCR process. DNA strands are separated and the whole procedure is repeated up to 20 times in order to select the best fitting sequences.

If RNA is used, a transcription step must be inserted before and after PCR. In order to increase specificity for the target molecule and exclude unspecific binding, counter selection steps can be employed. In those selection rounds no target is used and DNA strands with affinity to the support and container material are removed from the pool.

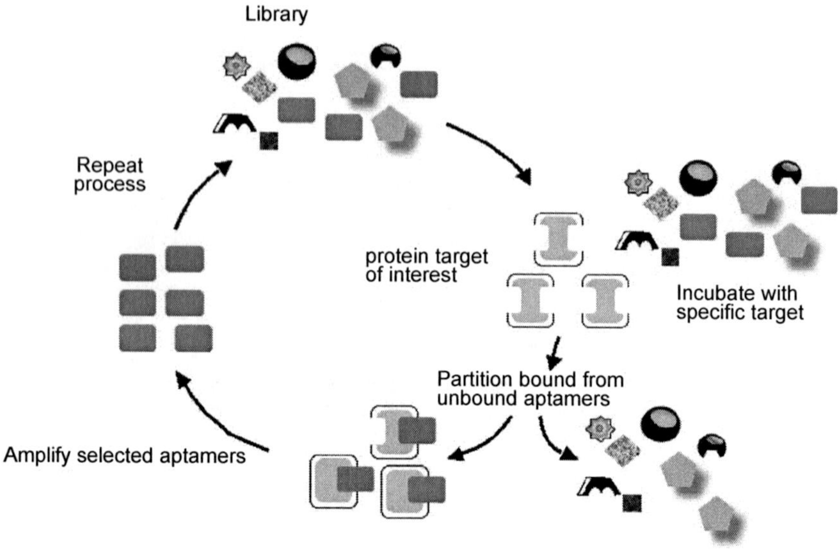

Figure 5.9: The SELEX process for use with a DNA library (Lee et al. 2008).

Biosensor Applications

In order to eliminate systematic problems with sandwich assays, the development of label free biosensors is an interesting topic. Xiao et al. (2005) modified a thrombin specific aptamer with a thiol group for immobilization on a gold surface. The strand was partially hybridized with a not fully complementary strand bearing a methylene blue (MB) tag.

In the presence of thrombin, the strands separated and the MB redox tag was approximated to the Au surface Fig. 5.10(a).

Similar signal-on detectors were developed by Baker et al. (2006) and Lai et al. (2006). The single stranded aptamer had the ability to hybridize with itself and form three loops upon target binding (see Fig. 5.10(b)). The conformational change brings the MB tag in proximity to the gold surface and allows for an electrical measurement. Both systems could be regenerated to a high degree, and thus are potentially reusable. Baker's system could detect cocaine concentrations as low as 500 μM in biological fluids even in the presence of contaminants.

So et al. (2005) attached thrombin binding aptamer to a single walled carbon nanotube (SWNT) which connected two electrodes.

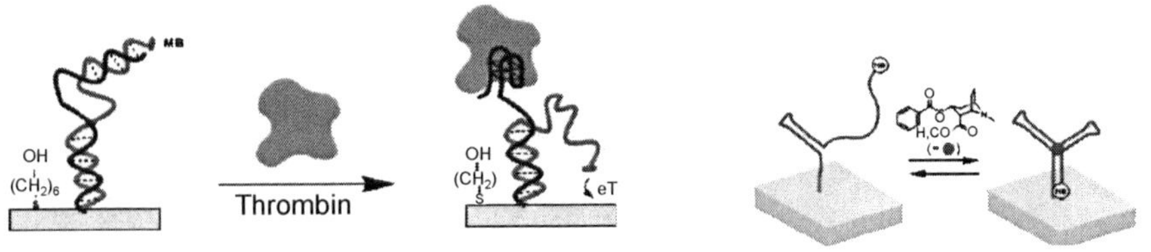

Figure 5.10: Aptamers with a methylene blue redox tag for thrombin (a) and cocaine detection (b). The binding event induces a conformational change in the aptamer and brings the redox active tag closer to the gold surface Baker et al. (2006); Lai et al. (2006). An analogue sensor was described earlier in Section 3.2. Xie et al. (2009) used carboxylic acid modified PEDOT nanowires instead of SWNTs as FET.

Color image of this figure appears in the color plate section at the end of the book.

Binding the charged protein induced an electrostatic gate potential and changed the source-drain current. The field effect transistor (FET) biosensor was able to detect thrombin in a concentration range of 10–100 nM.

OUTLOOK

In recent years, a fascinating development in the application of conductive polymers in medical diagnostics occurred. Elegant research on new sensing concepts has opened the door to a wide variety of microsystem based biosensors for clinical applications. Such devices are extremely useful for delivering diagnostic information in a fast, simple, and low cost fashion, and are thus uniquely qualified for meeting the demands of point of care systems, e.g., for cancer screening. The high sensitivity of the modern biosensors should facilitate early detection and treatment of diseases, and lead to increased patient survival rates.

In the future, one of the main challenges is to bring the new biosensor techniques to the bedside for use by non-laboratory personnel without compromising accuracy and reliability. The internal calibration and reference is also a major requirement, and provokes researchers to reshape the existing methods. From a clinical point of view, the *in vivo* biosensors that are biocompatible and can remain in the body for weeks or months will also be a demand. Special attention should be given to non-specific adsorption issues that commonly control the detection limits of electrochemical bioaffinity assays. The stability of biosensors remains an important issue in the fabrication and use of these devices for many application areas.

By measuring abnormalities within few minutes, disposable cartridges containing electrode strips and simple sample processing could offer early and fast screening of diseases in a point of care setting.

It has become apparent that the field of polymer biosensors has reached a new level of maturity. In the near future it is highly likely that pathogen detection will undoubtedly benefit from the integration of biosensors into all-polymer micro devices, and thus in some regards revolutionize the medical diagnostics.

ACKNOWLEDGEMENTS

This work was supported by the Danish Research Council for Technology and Production Sciences and the Technical University of Denmark.

REFERENCES

Akoudad, S. and J. Roncali. 2000. Modification of the electrochemical and electronic properties of electrogenerated poly(3,4-ethylene-dioxythiophene) by hydroxymethyl and oligo(oxyethylene) substituents. Electrochemistry Communications 2(1): 72.

Ali, E.M. and E.A.B. Kantchev, H-H. Yu and J.Y. Ying. 2007. Conductivity shift of polyethylenedioxythiophenes in aqueous solutions from side-chain charge perturbation. Macromolecules 40(17): 6025–6027.

Baker, B.R. and R.Y. Lai, M.S. Wood, E.H. Doctor, A.J. Heeger and K.W. Plaxco. 2006. An electronic, aptamer-based small-molecule sensor for the rapid, label-free detection of cocaine in adulterated samples and biological fluids. J Am Chem Soc 128(10): 3138–9.

Balamurugan, A. and S.-M. Chen. 2007. Poly(3,4-ethylenedioxythiophene-co-(5-amino-2-naphthalenesulfonic acid)) (pedot-pans) film modified glassy carbon electrode for selective detection of dopamine in the presence of ascorbic acid and uric acid. Anal Chim Acta 596(1): 92–8.

Balamurugan, S. and A. Obubuafo, S.A. Soper and D.A. Spivak. 2008. Surface immobilization methods for aptamer diagnostic applications. Anal Bioanal Chem 390(4): 1009–21.

Bhagat, A.A.S. and S.S. Kuntaegowdanahalli and I. Papautsky. 2008. Continuous particle separation in spiral microchannels using dean flows and differential migration. Lab on a Chip 8(11): 1906–1914.

Bhagat, A.A.S. and H. Bow, H.W. Hou, S.J. Tan, J. Han and C.T. Lim. 2010. Microfluidics for cell separation. Medical & Biological Engineering & Computing 48(10, Sp. Iss. SI): 999–1014.

Biswas, M. and A. Roy. 1994. Thermal, stability, morphological, and conductivity characteristics of polypyrrole prepared in aqueous medium. Journal of Applied Polymer Science 51(9): 1575.

Campbell, T.E. and A.J. Hodgson and G.G. Wallace. 1999. Incorporation of erythrocytes into polypyrrole to form the basis of a biosensor to screen for rhesus (d) blood groups and rhesus (d) antibodies. Electroanalysis 11(4): 215.

Daugaard, A.E. and S. Hvilsted, T.S. Hansen and N.B. Larsen. 2008. Conductive polymer functionalization by click chemistry. Macromolecules 41(12): 4321.

Di Carlo, D. and D. Irimia, R. Tompkins and M. Toner. 2007. Continuous inertial focusing, ordering, and separation of particles in microchannels. PNAS 104(48): 18892–18897.

Dubois, M-P. and C. Gondran, O. Renaudet, P. Dumy, H. Driguez, S. Fort and S. Cosnier. 2005. Electrochemical detection of arachis hypogaea (peanut) agglutinin binding to monovalent and clustered lactosyl motifs immobilized on a polypyrrole film. Chem Commun (Camb) (34): 4318–20.

Giddings, J. 1993. Field-flow fractionation: analysis of macromolecular, colloidal, and particulate materials. Science 260(5113): 1456–465.

Gossett, D.R. and W.M. Weaver, A.J. Mach, S.C. Hur, H.T.K. Tse, W. Lee, H. Amini and D. Di Carlo. 2010. Label-free cell separation and sorting in microfluidic systems. Analytical and Bioanalytical Chemistry 397(8): 3249–3267.

Gregoratto, I. and C. McNeil and M. Reeks. 2007. Micro-devices for rapid continuous separation of suspensions for use in micro-total-analysis-systems. SPIE.

Groenendaal, L. and F. Jonas, D. Freitag, H. Pielartzik and J.R. Reynolds. 2000. Poly(3,4-ethylenedioxythiophene) and its derivatives: Past, present, and future. Advanced Materials 12(7): 481.

Gustafsson, J. and B. Liedberg and O. Inganäs. 1994. In situ spectroscopic investigations of electrochromism and ion transport in a poly (3,4-ethylenedioxythiophene) electrode in a solid state electrochemical cell. Solid State Ionics 69(2): 145.

Han, K. and Z. Liang and N. Zhou. 2010. Design Strategies for Aptamer-Based Biosensors. Sensors 10(5): 4541–4557.

Hansen, T. and K. West, O. Hassager and N. Larsen. 2007. Direct fast patterning of conductive polymers using agarose stamping, Advanced Materials 19(20): 3261.

Jagur-Grodzinski, J. 2002. Electronically conductive polymers. Polymers for Advanced Technologies 13(9): 615.

Kim, D-J. and N-E. Lee, J-S.Park, I-J. Park, J-G. Kim and H.J. Cho. 2010. Organic electrochemical transistor based immunosensor for prostate specific antigen (psa) detection using gold nanoparticles for signal amplification., Biosens Bioelectron 25(11): 2477–82.

Kolb, H.C. and M.G. Finn and K.B. Sharpless. 2001. Click Chemistry: Diverse Chemical Function from a Few Good Reactions. Angewandte Chemie International Edition 40(11): 2004.

Kumar, S.S. and J. Mathiyarasu, K.L.N. Phani and V. Yegnaraman. 2006. Simultaneous determination of dopamine and ascorbic acid on poly (3,4-ethylenedioxythiophene) modified glassy carbon electrode. Journal of Solid State Electrochemistry 10(11): 905.

Kuntaegowdanahalli, S.S. and A.A.S. Bhagat, G. Kumar and I. Papautsky. 2009. Inertial microfluidics for continuous particle separation in spiral microchannels, Lab on a Chip 9(20): 2973–2980.

Lai, R.Y. and D.S. Seferos, A.J. Heeger, G.C. Bazan and K.W. Plaxco. 2006. Comparison of the signaling and stability of electrochemical dna sensors fabricated from 6- or 11-carbon self-assembled monolayers. Langmuir 22(25): 10796–800.

Lazcka, O. and F.J.D. Campo and F.X. Muñoz. 2007. Pathogen detection: A perspective of traditional methods and biosensors, Biosensors and Bioelectronics 22(7): 1205–1217.

Lee, J.-O. and H-M. So, E-K. Jeon, H. Chang, K. Won and Y.H. Kim. 2008. Aptamers as molecular recognition elements for electrical nanobiosensors. Anal Bioanal Chem 390(4): 1023–32.

Lee, M-W. and M-Y. Lee, J-C. Choi, J-S. Park and C-K. Song. 2010. Fine patterning of glycerol-doped pedot:pss on hydrophobic pvp dielectric with ink jet for source and drain electrode of otfts. Organic Electronics 11(5): 854–859.

Lenshof, A. and T. Laurell. 2010. Continuous separation of cells and particles in microfluidic systems. Chemical Society Reviews 39(3): 1203–1217.

Loffredo, F. and A.D.G.D. Mauro, G. Burrasca, V. La Ferrara, L. Quercia, E. Massera, G. Di Francia and D.D. Sala. 2009. Ink-jet printing technique in polymer/carbon black sensing device fabrication. Sensors and Actuators B: Chemical 143(1): 421.

Mabrook, M. and C. Pearson and M. Petty. 2006. Inkjet-printed polypyrrole thin films for vapour sensing. Sensors and Actuators B: Chemical 115(1): 547.

Nolan, J. and L. Sklar. 1998. The emergence of flow cytometry for sensitive, real-time measurements of molecular interactions. Nat Biotechnol 16(7): 633–8.

Olson, T.S. and S. Pylypenko, P. Atanassov, K. Asazawa, K. Yamada and H.Tanaka. 2010. Anion-exchange membrane fuel cells: Dual-site mechanism of oxygen reduction reaction in alkaline media on cobalt–polypyrrole electrocatalysts. The Journal of Physical Chemistry C 114(11): 5049.

Park, J. and H.K. Kim and Y. Son. 2008. Glucose biosensor constructed from capped conducting microtubules of pedot. Sensors and Actuators B: Chemical 133(1): 244.

Peterson, F. and L. Åberg, A. Swärd-Nilsson and T. Laurell. 2007. Free Flow Acoustophoresis: Microfluidic-Based Mode of Particle and Cell Separatio. Anal Chem 79(14): 5117–5123.

Pirrung, M.C. 2002. How to make a DNA chip. Angew. Chem. Int Ed Engl 41(8): 1276–89.

Prodromidis, M. and M. Karayannis. 2002. Enzyme based amperometric biosensors for food analysis. Electroanalysis 14(4): 241.

Rozlosnik, N. 2009. New directions in medical biosensors employing poly(3,4-ethylenedioxy thiophene) derivative-based electrodes. Analytical and Bioanalytical Chemistry 395(3): 637–45.

Sarma, A.K. and P. Vatsyayan, P. Goswami and S.D. Minteer. 2009. Recent advances in material science for developing enzyme electrodes. Biosens Bioelectron 24(8): 2313–22.

Seo, J. and M. Lean and A. Kole. 2007. Membrane-free micro-filtration by asymmetric inertial migration. Appl Phys Lett 91(033901).

Shirakawa, H. and E.J. Louis, A.G. MacDiarmid, C.K. Chiang and A.J. Heeger. 1977. Synthesis of electrically conducting organic polymers: halogen

derivatives of polyacetylene, (CH) x . Journal of the Chemical Society, Chemical Communications (16): 578.
So, H.-M. and K. Won, Y.H. Kim, B.-K. Kim, B.H. Ryu, P.S. Na, H. Kim and J-O. Lee. 2005. Single-walled carbon nanotube biosensors using aptamers as molecular recognition elements. J Am Chem Soc 127(34): 11906–7.
Syed, M.A. and S. Pervaiz. 2010. Advances in Aptamers. Oligonucleotides 20(5): 215–224.
Teles, F. and L. Fonseca. 2008. Applications of polymers for biomolecule immobilization in electrochemical biosensors. Materials Science and Engineering: C 28(8): 1530.
Toner, M. and D. Irimia. 2005. Blood on a Chip. Annu. Rev. Biomed Eng. pp. 77–103.
Tuerk, C. and L. Gold. 1990. Systematic evolution of ligands by exponential enrichment: Rna ligands to bacteriophage t4 dna polymerase. Science 249(4968): 505.
Vasantha, V. and S. Chen. 2006. Electrocatalysis and simultaneous detection of dopamine and ascorbic acid using poly(3,4-ethylenedioxy)thiophene film modified electrodes. Journal of Electroanalytical Chemistry 592(1): 77.
Wang, L-X. and X-G.Li and Y-L. Yang. 2001. Preparation, properties and applications of polypyrroles. Reactive and Functional Polymers 47(2): 125–139.
Wang, Y. and A. Stevens and J. Han. 2005. Million-fold preconcentration of proteins and peptides by nanofluidic filter. Analytical Chemistry 77(14): 4293–4299.
Wessling, B. 2001. From conductive polymers to organic metals. Chemical Innovation USA 31(1): 34–40.
Xiao, Y. and B.D. Piorek, K.W. Plaxco and A.J. Heeger. 2005. A reagentless signal-on architecture for electronic, aptamer-based sensors via target-induced strand displacement. J Am Chem Soc 127(51): 17990–1.
Xie, H. and S-C. Luo and H-H. Yu. 2009. Electric-field-assisted growth of functionalized poly(3,4-ethylenedioxythiophene) nanowires for label-free protein detection. Small 5(22): 2611–7.
Zimmermann, G.R. and C.L. Wick, T.P. Shields, R.D. Jenison and A. Pardi. 2000. Molecular interactions and metal binding in the theophylline-binding core of an rna aptamer, RNA-A Publication of the RNA Society 6(5): 659–667.

6

Biosensors for Diagnostic Based on Olfactory Receptors

Marta Sanmartí,[1] *Patrizia Iavicoli*[2] and *Josep Samitier*[3,*]

CONCEPT OF BIOELECTRONIC NOSES

Introduction

An electronic nose is a sensing device capable of producing a digital fingerprint, also known as pattern, of a group of chemical volatiles which characterize a specific odour. Such devices consist of two main components: a chemical sensing system and a pattern recognition system (Ortega et al. 2000).

The first reported developments about sensor noses are dated to the 1970s.The possibility of using redox reactions of volatile compounds and detecting the conductivity change was studied. Sensor technology has developed rapidly over the last three decades and this has resulted in a range of different sensor formats and in the development of complex microarray sensor devices. In the specific area of electronic-nose systems, several different physicochemical techniques have been used to produce

[1]Institute for Bioengineering of Catalonia (IBEC), c/Baldiri Reixac 10-12, 08028 Barcelona, Spain.

[2]Centro de Investigación Biomédica en Red. Bioingeniería, Biomateriales y Nanomedicina (Ciber-bbn).

[3]Department of Electronics, University of Barcelona, c/Martí i Franquès 1, 08028 Barcelona, Spain; E-mail: jsamitier@ub.edu

*Corresponding author

sensor arrays for odour characterization. Since 2000 more than 1800 papers have been published with the label electronic nose (ISI web of knowledge) with annual production between 180 and 200 in the last 5 yr.

The first complete electronic nose system was reported by Persaud and Dodd in 1982. It was based on the idea of mimicing the human olfactory system to detect different volatile compounds by using biochemical sensors. The sensor signal responses treatment was done using neural networks to recognize the key components presents in the odour.

Technologies used for electronic nose sensors include the main transducer mechanisms used to obtain general chemical sensors (Gardner and Bartlett 1999). Depending of the transducer mechanism we can distinguish between:

Conductivity sensors based on polymers or metal oxides. In this case the interaction between the chemical volatile and the material produce a modification of the conductivity. The selectivity depends on the material, which can be doped, and on the temperature conditions. Some drawbacks to be considered in the design are the dependence on humidity and the drift over time.

Field-effect sensors. These sensors are related to metal oxide sensors but the output signal is obtained from a change in potential when the volatile molecules react at a catalytic surface.

Piezoelectric sensors. They use the radio frequency resonance of quartz materials coated with acetyl cellulose or lecithin membranes. The adsorption of volatile molecules onto the membrane produces a change in the magnitude of the resonance frequency that is related to the mass of the volatile analyte. The selectivity of these sensors depends on the thickness of the coatings.

Surface acoustic-wave devices. These devices are based on waves that are emitted along the surface of a crystal by the electric field of surface-deposited aluminium electrodes.

Optical sensors. These sensors are based on a light source that excites the volatile analyte, and the signal can be measured in the resulting absorbance, reflectance, fluorescence or chemiluminescence. In the case of surface plasmon resonance devices, changes in the optical properties of the polymer materials can be also used for voltalie detection.

Electrochemical sensors. These sensors contain electrodes and an electrolyte. The responses generated are dependent on the electrochemical characteristics of the volatile molecules that are oxidized or reduced at the working electrode, with the opposite occurring at the counter electrode.

The voltage or current generated by the reactions between the electrodes is measured.

There are a large number of papers about all these sensor technologies based on chemical sensors (Pearce et al. 2002), but the main drawback is the general poor selectivity, which requires a complex signal processing post-treatment. In many reported papers the devices obtained can be used only in specific environments where the number of volatiles and the ambiance conditions can be limited and controlled. For a more detailed view of the last technological developments see: Wilson and Baietto (2009) and (2011); Lee and Park (2010).

Bioelectronic Noses

The main focus of this chapter is to analyze the olfactory system in mammals and insects and review the last technological development of bioelectronic noses based on the use of biological receptors and the ability to be used as diagnostic systems of different diseases.

The development of a general purpose bioelectronic nose system which mimic the physiological olfaction system of insects or mammalians constitute a promising scientific and technological research field (Turner 2000). The mammalian nose is the gold standard of odour detection, displaying both an unmatched chemical space (estimates vary from tens of thousands up to one million of compounds) and the highest sensitivity. These performances are due to both the primary olfactory message, which is initially elaborated by the olfactory neurons, and the subsequent neuronal processing. The olfactory receptor neurons are one of the main classes of the chemoreceptors. As the chemoreceptor is the known sensory receptor that transduces a chemical signal (stimuli in the environment) into an action potential, and is also known as chemosensor. In comparison with the ear and eye, the mammalian nose is much more complicated, at least regarding the mechanisms responsible for the primary reaction to an external stimulus. In olfaction hundreds of different classes of biological receptors are involved, whereas in vision in principle only three different classes are found. The human olfactory system is very complex, and is not yet fully understood.

VOLATILES DETECTION FOR DIAGNOSTIC OF DIFFERENT DISEASES

Odours emitted by the human body and present in body fluids characterize an individual, and depend both on its genetic signature and on its physio-pathological status. Some medical practices have used, or are still using, sensorial diagnosis (mainly olfactory) to check good health or to determine various pathologies (phenylcetonuria, typhoïd, scorbut, etc.).

Gas Chromatography and Gas Chromatography-Mass Spectroscopy techniques have already been used to monitor patterns of volatile compounds as an aid to the clinical diagnosis of aerobic and anaerobic bacterial infections and cardiopulmonary disease (Fiñipiak 2010).

The patient samples used include human pus, urine, blood plasma and alveolar air. However, the uses of these techniques require expensive analytical equipment and are time consuming. Perhaps in the future, it might be possible to have electronic nose devices as point-of-care systems that could be used as a rapid screen for specific diseases or disorders (Anthony et al. 2004).

Table 6.1: Schematic comparison between human and electronic noses (Adapted from Davide et al. 2001).

MAMMALIAN OLFACTION	ELECTRONIC NOSE based on Chemical Sensors
Around 10–40 millions	5–100 chemical receptors
Around 100 to 1000 selectivity classes	5–100 selectivity patterns
Initial reduction of number of signals From 1000 to 1	Smart signal processing
Cannot detect very small molecules	Can detect simple molecules (H_2; CO_2, CO)
Limit detection 0,1 ppb (Insects 10^{-6} ppb)	Limit detection around 0,1 ppb
Size Olfactory neuron cell 40–50 microns	Size of chemical sensors 100–800 microns
Size olfactory receptor 5 nanometers	Size of nanosensors around 100nm

As an example, studies *in vitro* have shown that it is possible to discriminate between different aerobic bacteria, such as *Helicobacter pylori*, *Escherichia coli* and *Enterococcus* species that are present in samples, both alone and as a mixture of the three species, on the basis of differences in the amounts of terpenes, trimethylamine and ketones produced (Pavlou et al. 2000) *Mycobacterium tuberculosis*, the causative agent of tuberculosis, has been detected in cultured sputum samples either directly or following treatment with enzymes to enhance volatile production. Samples from patients with urinary tract infections (UTIs) and tuberculosis were identified correctly in 90–99% of cases compared with traditional culture techniques (Pavlou 2002).

Electronic-nose devices can be also used to detect diseases other than infectious diseases. For instance; it has been shown that patients with kidney disorders produce characteristic volatile compounds, which can be a useful tool in the diagnosis and control of renal dialysis (Di Natale et al. 1999). Additionally, studies by the same group have shown that lung cancer can be detected by breath analysis using non-selective gas sensors (Di Natale et al. 2003).

Recently, Peng et al. (2010) have shown that volatile organic compounds (VOC) analysis of exhaled breath using an array of nanosensors can be useful to detect lung, breast, colorectal and prostate cancers.

A major challenge of these diagnostic tools lies in the establishment of a precise odourant disease signature, understood as the ensemble of volatile compounds whose concentration in a body fluid (i.e., urine, blood, pus, etc.) or in the breath varies in patients with the malignancy with respect to healthy individuals. For bladder and prostate cancers formaldehyde has already been shown to be one member of the odourant disease signature of the patients' urine (Spanel et al. 1999).

Among their medical applications, the detection of bacterial pathogens or pneumonia, and the monitoring of glucose levels and detection of lung cancer by polymer composite sensor array analyses of exhaled breath has been reported (Hanson and Steinberger 1997, Dalton et al. 2004, Machado et al. 2005, Chen et al. 2005).

Animals' Olfaction as a Biological Recognition Element

As compared to traditional (or even up-to-date) electronic noses, animal olfaction is repeatedly reported to display much lower detection limits (below 0.1 ppb down to 10^{-6} ppb, or even less) in some insects (Angioy et al. 2003).

In association with this unchallenged sensitivity, animal olfaction offers a second outstanding analytical property, that is, the capability of detecting a wealth of odourant chemicals, provided that they are of rather small molecular size (molecular weight below 300 dalton). This is made possible thanks to the so-called 'combinatorial code of odours', through which the repertoire of olfactory receptors (about 1500 genes in rodents), combined with the possibility for each given receptor to detect (and likely discriminate, at least partially) 10s of odourants (Malnic et al. 1999) yields an amazing number of different possibilities to identify chemical compounds. Most of these characteristics of animal olfaction arise directly from the olfactory receptors themselves.

Use of Dog Olfaction Abilities for Diagnostic Purposes

The use of dogs for diagnostic purposes of skin cancers was published for the first time as anecdotal reports by Williams and Pembroke (1989), Church and Williams (2001). In 2004, Willis et al. reported that dogs can distinguish urine from patients with bladder cancer with a mean success rate of 41% and Elliker et al. (2004) show the capability of dogs to detect prostate cancer. Confirmed this year by Cornu et al. (2011) obtaining a sensitivity of 0, 91 and a specificity of 0.91.

More recently, McCulloch et al. (2006), show that ordinary household dogs can be trained to distinguish breath samples of patients with lung and breast cancer with high accuracy (sensitivity and specificity of 0,99 and 0,99 in lung cancer and 0,88 and 0,98 in breast cancer respectively). Horvath and coauthors (2008) reported that between ovarian cancer tissues samples and control tissues, the sensitivity obtained was 100% and the specificity 97,5%.

This year 2011, Sonoda et al., report an exhaustive analysis of colorectal cancer screening performed by dogs using breath samples of patients and compared with colonoscopy showing a good sensitivity and specificity (0,91 and 0,99 respectively). Most interesting is that accuracy of canine scent detection was even higher for early stage cancers and canine scent detection was not confounded by current smoking, benign colorectal disease, inflammatory disease or the presence of human haemoglobin or transferin. The authors conclude that the results indicate that a specific cancer scent indeed exists. The volatile organic compounds detected during canine scent presumably occur early in the pathogenesis of colorectal cancer.

HOW TO UNDERSTAND THE OLFACTORY SYSTEM

The Olfaction in Vertebrates and Insects

The odours detection allows the animals to locate food, danger, to avoid predators, to identify mates, to communicate between themselves and to reproduce. The animals in their natural environments are immersed in odours, the detection of which is necessary for their survival. The odourants, these chemical substances, are rich in information, for that reason the animals have evolved to detect and interpret them. Depending on the biological information that they give are known as odourants or pheromones.

When we talk about the sense of smell, this may involve a large number of distinct receptors capable of associating with one or a small number of odourants (Buck and Axel 1991). In general, olfactory systems have evolved great sensitivity and discriminatory power. It is reported in the literature that the mammalian olfactory system can recognize and discriminate a large number of different odourant molecules. The detection of the odourants results from the interaction between odourant ligands and specific receptors on olfactory sensory neurons (OSNs). The identification of a large family of G-protein-coupled receptors (GPCRs) in the nose manages the olfactory system discriminatory task (Firestein, 2001).

However, several recent reviews claim that odourant receptors and olfactory transduction in vertebrates and insects are different. Otherwise, in both insects and mammals, odourants binds to receptors in the cilia or dendrites of olfactory receptor neurons (ORNs) (Su et al. 2009).

The Vertebrate Olfaction

It is reported that the olfaction takes place in several olfactory subsystems of the nose. During the last 20 years the following five types of chemosensory GPCRs in the mammalian nose have been identified: odourant receptors (ORs) (Buck and Axel 1991), trace amine-associated receptors (TAARs) (Liberles and Buck 2006), two distinct vomeronasal receptors (V1R and V2R) (Dulac and Axel 1995, Matsunami and Buck 1997), and formyl peptide receptors (FPRs) (Riviere et al. 2009). Until now, the ORs are the largest family, with up to 2130 genes discovered. The ORs are expressed in the main olfactory epithelium of mammals and bind small, volatile molecules as it has been mentioned before (Kaupp 2010). So, the ORs are responsible for the sense of smell. Vertebrates ORs are known to be expressed in different olfactory sensory neurons in the nasal neuroepithelium (Breer 2003, Buck and Axel 1991). The olfactory sensory neuron has the characteristic to be bipolar: each OSN extends a dendrite to the mucosal surface and an axon to the olfactory bulb of the brain which is involved in neural processing. When the axons reach the olfactory bulb, they make synapses with the dendrites of projection neurons, within discrete structures known as glomeruli. The characteristic of the dendrites is that they include the cilia which contain the ORs and their main goal is to increase the surface area of their exposure to volatiles.

In vertebrates, as it has proved in 2000s (Mombaerts and Nei 2004, Touhara and Vosshall 2009) each OSN express mainly a single type of OR, only one receptor gene. But in the nasal mucosa there are other proteins that have an important role. These proteins allow the OR-odourant interaction process, such as the water-soluble odourant-binding proteins (OBPs). It is difficult to prove the specific function of these proteins but there is some hypothesis that seems to focus that they act to solubilize and transport specific odourants, making them available to the ligand-binding site of ORs in order to enable the ORs in the aqueous phase, to access volatile compounds (Glatz and Bailey-Hill 2011).

The Insect Olfaction

Before 1999 it was reported that insects, as well as mammals, detect odours through GPCRs. Three laboratories identified a family of these seven-transmembrane receptors whose expression in the antennae of the fruit fly *Drosophila melanogaster* turned out to control the responses of individual olfactory neurons to chemical odourants (Chesler and Firestein 2008). The first experiments prove that insects detect odours differently from mammals came from the analysis of the receptors' amino-acid sequences. As well as mammals' olfaction system, the insects' one is divided in olfactory subsystems. In flies, the number of chemosensory systems is lower than in mammals. In the *D. melanogaster* three chemosensory subsystems have been identified: olfactory receptors (that are not related with the vertebrates ORs) localized in the basiconic and trichoid censilla, gustatory receptors (GRs) expressed in taste organs through the body and ionotropic glutamate receptors (IRs) expressed in coeloconic sensilla. The flies, with a difference of the vertebrate olfactory system, have two bilaterally symmetrical pairs of olfactory organs, the third segment of the antennae and the maxillary pals (located in the head).

The observation of Sato et al. (2008) gives evidence that for odour detection, insects use an unusual strategy and unusual receptor (they focus their studies on the fact that the receptor is an ion channel). But there are other groups, such as Wicher et al. (2008) that have different views to understand the sequence of events that occur from ligand binding to channel opening.

Volatile Molecules Detection by the Olfaction System The Role of the Olfactory Receptors

By studying odourant receptors we can understand the events of olfaction providing crucial knowledge for the development of a novel, odourant receptor-based biosensor for chemical screening. Each odourant receptor seems to recognize multiple odourants, as well as each odourant seems to be recognized by more than one odourant receptor. This combinational odourant receptor coding scheme allows the differentiation of a great number of several odours with relatively few odourant receptors.

Several studies conducted in odour detection of vertebrates supported a G-protein-coupled, cAMP-mediated transduction in the main olfactory system (Spehr and Munger 2009). These types of proteins, the GPCRs, are very well-known. The vertebrates OR genes encode a large family of GPCRs to recognize the odourant molecules in the olfactory ephitilium. In particularly, the mammalian OR gene family encodes around 400 and 1200 different olfactory receptor proteins in humans and mice respectively, indeed they are the largest family of genes reported in the literature (Zhuang and Matsunami 2008). The sequence of the GPCRs proteins contains seven segments with mostly hydrophobic amino acid residues that form an α-helix structure through the lipid bilayer membrane, alternated with hydrophilic segments that form loops, which are stable in water, connecting each helices (Breer H. 2003, Zarzo 2007,

DeMaria and Ngai 2010). The GPCRs are a group of proteins that exhibit a seven-transmembrane domain topology. They are very well-known proteins and have two important binding sites: the odourant binding site on the mucosal layer (extracellular side) and the binding site made up of three subunits on the cytoplasmatic domain that couples a heterotrimeric G-protein (guanine nucleotide binding protein).

When the odourant attaches to the binding site of the GPCR the olfactory signal transduction is initiated. The binding gives a conformational change that activates an olfactory-specific subtype of G-protein, the $G_{\alpha olf}$. This subunit is responsible for activating the adenylyl cyclise enzyme. The role of this enzyme in the olfactory system is to convert intracellular adenosine triphosphate (ATP) into cyclic adenosine monophosphate (cAMP), following with a cascade of events that lead to the nerve cell signal (Firestein 2001). This is the first step of the transduction cascade that is organized in two stages. After that, the cyclic AMP levels increase the opening nucleotide-gated channel, giving a influx of Na^+ and Ca^{2+} ions (Song et al. 2008) that will generate a potential done by Ca^{2+-}dependent Cl^- channels (the depolarization of the cell is amplified through the activation of these ions). After this process, the ORNs accumulate chloride through the $Na^+/K^+/2Cl^-$ cotransporter NKCC1 and maintain an elevated intracellular Cl^- concentration to support amplification. (Hengl et al. 2010)

Recent studies about olfactory receptors in insects have been published (Sato et al. 2008, Wicher et al. 2008). They proved that the co-receptor OR83b (expressed in all OSNs) together with the OR forms a heteromeric ligand-gated cation channel in the Drosophila. What is not clear at present is the exact mechanism of fly signal transduction. The receptors in insects recognize olfactory stimuli by ion channels, not only GPCRs. (Kaupp 2010).

One of the authors reported that evidence does not exist for the involvement of G-proteins or intracellular messengers such as cAMP,cGMP or inositol-1,4,5-trisphoshate for a fast ionotropic response (Sato et al. 2008). They suggest that an ion channel opened when an odourant binds the odourant receptor directly. In that case, the receptor consists of a common receptor (a GPCR) and a co-receptor (OR83b). This hypothesis doesn't specify the location of the channel pore and would be the simplest model. On the other hand, another group proved that common insect ORs activate the synthesis of cAMP through the G-protein, and thus this in turn activates OR83b co-receptor, which serves as a cAMP-gated ion channel (Wicher et al. 2008). They suggest two different pathways by which odour-induced-depolarization can be generated. When an odourant binds the OR, the signal is transferred to the OR83b co-receptor. This transfer can be direct or indirect. The direct transfer consists of opening a channel formed by the OR83b subunit directly after the odourant binding, giving a fast and short depolarization. However, the authors also consider an alternative, the indirect pathway. The activation of a G-protein and an adenylyl cyclise allows the production of the cAMP. When the cyclic AMP binds the co-receptor OR83b, the channel opens and generates slow and prolonged depolarizing currents. This controversy between the different studies done on invertebrates olfactory receptors is very well-documented in the literature. (Kaupp 2010).

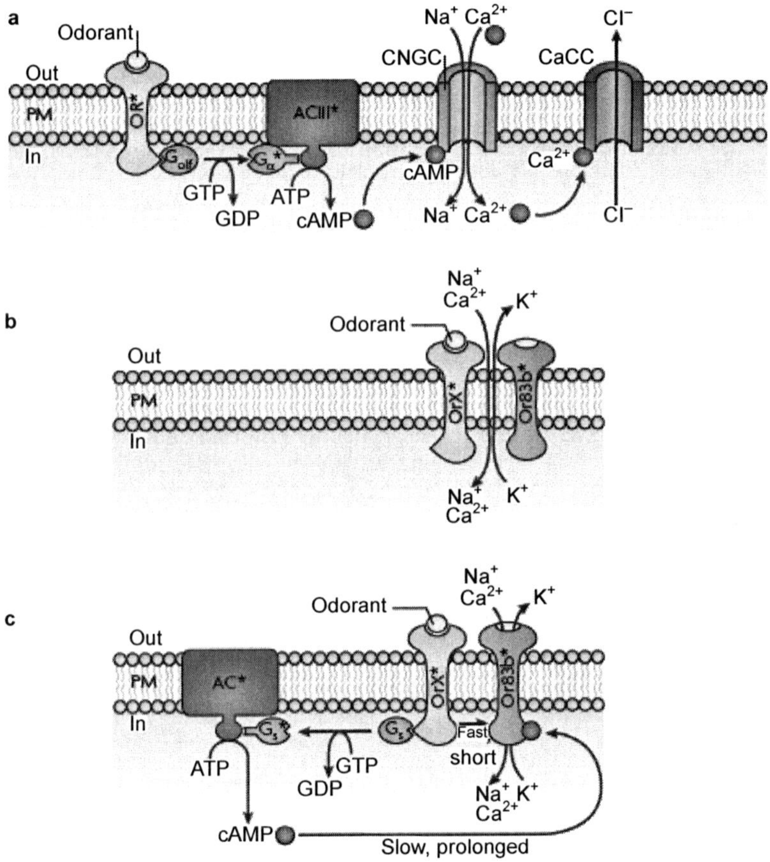

Figure 6.1: Adapted from Kaupp 2010. Nature Reviews. a) Odorant detection in vertebrates. When the odorant binds the olfactory receptor (OR) activates the trimeric transmembrane protein, olfaction-specific G protein (Golf), adenylyl cyclase type III (ACIII), the olfactory cyclic nucleotide-gated channel and a Ca2+-activated Cl - channel (CaCC). b) Insect olfactory model suggests that the odorant receptor forms an ion channel that is opened directly in response to the binding of odorants. c) On the other hand this complex insect model suggests that there are two pathways by which odour-induced depolarization can be generated. Upon odorant binding, activity is transferred to the Or83b subunit either by a direct (fast and short) or indirect (slow and prolonged) pathway.

Color image of this figure appears in the color plate section at the end of the book.

The Role of odourant-binding Protein

When odourants enter the nasal cavity two types of processes occur. The important one is the specific interaction with receptor proteins on the membrane of the olfactory cilia as it has been presented until now. But, it should not be forgotten that before reaching the neuronal membrane, the odourant molecules have to cross a layer of mucus containing high concentration of several classes of proteins that will interact with the odourants, one of these proteins are the odourant-binding proteins, that are extracellular proteins located in the fluid surrounding the sensory dendrite. Like other similar systems, the olfactory mucus is very complex in its composition and several aspects are under investigation.

However it is known that odourant-binding proteins (OBPs) play a crucial role in the interaction between the OR and the volatile compound. Till now, the complete role of OBPs in olfactory biosensors of their physiological function remains unresolved. For this reason some OR signalling *in vitro* studies include the OBP and in others do not (Glatz 2011). However, odour solubilization is still required in these researches. In the last years, some groups have focused their studies on the role of these proteins and how to improve their previous results. For example, Vidic et al. (2008) investigated the role of OBP-1F in regulating binding kinetics of the ligand helional on yeast-derived nanovesicles, nanosomes, containing OR17-40. They showed that the presence of OBP-1F increased sensitivity at lower odourant concentrations. Assuming that the sensitivity is a highly important issue in the development of an effective biosensor, Park et al. (2010) also proved that OBPs can be used to increase the sensitivity of olfactory receptor-based biosensors. Their results suggest that OBPs can bind to different olfactory receptors with different specificity and have the highest affinity to olfactory receptors that originate from the same species (Ko et al. 2010). Thus, they can bind to various different odourants.

It is well-known that odourant-binding proteins are low-molecular-weight proteins highly concentrated in the nasal mucus of vertebrates and in the sensillar lymph of insects with affinity towards odours and pheromones. They were discovered by Pelosi during the 80s in the search for olfactory receptors using radioactively labelled odourants in ligand-binding experiments, and were detected in the nasal olfactory mucosa of bovines and other mammals as a soluble protein able to fix pyrazine (2-isobutyl-3-metoxypyrazine) with low detection thresholds. They hypothesized that this protein could serve as an odourant-carrier. (Tegoni 2000). However if the different species are compared, literature shows no homology between vertebrate OBPs and insect OBPs in amino acid sequence and three-dimensional structure (Pelosi et al. 2005, Zhou 2010).

If we focus on the vertebrates OBPs, these proteins are small lipocalin-like proteins defined by their property of reversibly binding volatile chemicals, but insect OBPs do not have this structural morphology (Pelosi 1994, Vosshall and Stensmyr 2005, Glatzand and Bailey-Hill 2011). As reported in the literature, insects OBPs are better characterized than mammalian ones. The first study of the OBPs in insects was in the 80s that were discovered as small, secreted molecules present in fluid that bathed pheromone-sensitive OSNs and originally called pheromone-binding-proteins. In addition, the first OBP of insects (called pheromone-binding protein, PBP) was identified in the moth *Antheraea polyphemus* using its sex-specific pheromone (Vogt and Riddiford 1981).

Till now, a lot of OBPs have been discovered and classified in three groups: pheromone-binding-protein, general odourant-binding proteins groups 1 and 2 (Wang et al. 2003).

Our main goal is to understand how the mammalian olfactory system works, for this reason we will focus on the vertebrates OBPs. Some structure analysis reported in the literature revealed that generally the OBPs consist of a polypeptide chain of about 150 amino acids. This globular protein is folded into eight antiparallel b-sheets with α-helical domain located near the carboxy terminal. The β-sheets form a continuously hydrogen-bonded β-barrel, with the internal cavity accommodating the hydrophobic small ligands (Breer 2003). The binding pocket located inside the barrel can accommodate ligands of 10–20 carbon atoms of hydrophobic nature. Also, seems that each OBP has a unique ligand-binding profile. According to this hypothesis, some authors present the role of the OBPs as key proteins involved in preselecting those volatile compounds that finally interact with the olfactory sensory cells. In other words, the OBPs could play a role as a specific filter rather than a passive carrier protein for odourants in the mucus layer.

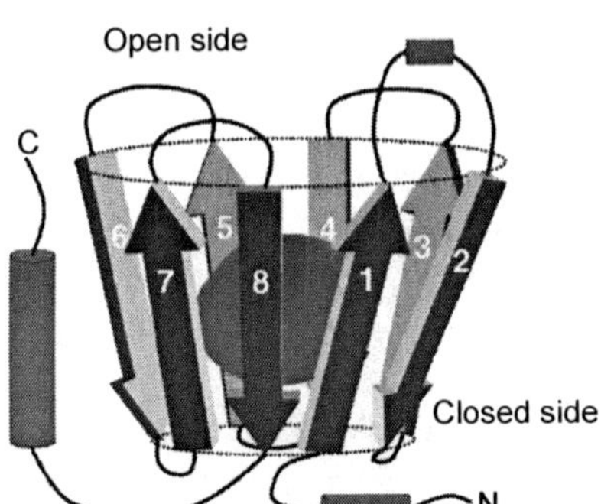

Figure 6.2: Schematic representation of an odorant-binding protein (Breer 2003).

Color image of this figure appears in the color plate section at the end of the book.

Volatile Molecules: An Approach from Odourants to the Pheromones Kingdom

As it was mentioned previously, odour detection is crucial for animal survival. These small molecules, called odourants, are defined as a volatile chemical compound with a molecular weight lower than 300 Da that humans or other animals can perceive. It is important not to confuse odourants with pheromones. Taking this into account we can define an odourant as any molecule capable of stimulating the olfactory neurons.

The exact number of odourant molecules on earth is unknown. However, literature has reported that humans can distinguish and detect approximately 10000 different odours. However, the exact number of known smells or the number that we can perceive is unknown (Touhara and Vosshall 2009).

How we can classify the volatile substances? A substance is volatile depending on the strength of intermolecular forces. For example, charged compounds are odourless because volatility is minimal when strong ionic or polar interaction is present (Pelosi 1994). However, these characteristics are not common between all animals. For example, the best odourants for aquatic animals are amino acids. These days it seems that irrespective of the structure or chemical properties, an odourant is volatile enough to reach the nose.

On the other hand, pheromones, defined as specific substances that are secreted by an individual and received by a second individual from the same species, or induce a specific reaction such as stereotype behaviour or endocrinological change (Touhara and Vosshall 2009). It is reported that the same pheromone can have a variety of effects, depending on the context or the receiver (Wyatt 2010).

Sometimes, the difference between a general odourant and a pheromone is not clear. For this reason it is easier to define a pheromone as a substance that is utilized for intraspecies communication. Pheromones themselves can send information about sex, strain, and be meaningful or informative for the species, mediate an amazing diversity of animal behaviour. The first result describing the pheromone kingdom was about pheromone moths, how these animals were capable of detecting a single molecule of female pheromone over an enormous distance.

During these years a lot of work have been done trying to understand the functional properties of these neurons (capable of detecting the pheromones) and the chemistry of these chemicals, but until now no evidence can clearly explain the biochemical mechanisms by which pheromones selectively activate sensory neurons (Vosshall and Stensmyr 2005). The vomeronasal organ (VNO) is the one that detects the pheromones. This is located in the nasal cavity and is a chemosensory system different from the olfactory sensory organ. Two families of VN receptor genes that encode proteins with seven transmembrane domains, V1R and V2R have been identified in the VNO.

Most of the mammal olfactory system have two olfactory organs: the main olfactory epithelium (MOE) and the vomeronasal organ (VNO), then the information is proceeded in separate neural pathways.

Despite their differences, invertebrates and vertebrates, in a wide range of habitats, are similar to each other in the way that they use chemical communication.

BIOELECTRONICS NOSES BASED ON *IN VIVO* EXPERIMENTS (WHOLE ANIMAL AND TISSUE-BASED RECOGNITION ELEMENT)

More than one million species of insects with sometimes extraordinary sensory abilities present a tremendous potential of highly optimized chemoreceptors. In order to make these abilities usable for mankind in terms of analytical tools, some interface between chemoreceptive organs of insects and microelectronic components of analytical instruments has to be established. One promising strategy is the design of biosensors on the basis of intact chemoreceptors utilizing electrophysiological techniques like electroantennography (EAG), which records the 'slow' changes of potential that are caused by the superposition of simultaneous membrane depolarizations of numerous receptor cells. Considering that detection limits of analytes are typically in the ppb range or even lower, a biosensor on the basis of intact chemoreceptors could serve as an analytical device with unrivalled data acquisition time. Therefore, the number of applications of this biosensor type is increasing. Most of them are developed for agricultural problems in integrated pest management as detection of insect pheromones, e.g., in greenhouses, in cotton fields (Baker and Haynes 1989, Sauer et al. 1992) or for the detection of plant damage in potato fields for the timing of pesticide application (Schütz et al. 1996).

In order to circumvent major drawbacks of conventional EAG methods, such as electrical and mechanical instability, the need for preamplification and the limited ability for miniaturization, new methods were designed. A biosensor device based on a field-effect transistor (FET)–insect antenna junction was presented for the first time by Shoning et al. (1998).

An antenna of the Colorado potato beetle (*Leptinotarsa decemlineata Say*) was directly connected by means of an electrolyte solution with the gate of a FET device which acted as a measuring electrode. When an air current loaded with a component of host plant odour (Z-3-hexen-1-ol) was flushed over the antenna, the induced depolarization response would modify the conductance of the FET channel between source and drain.

This resulted in a corresponding change of the drain current as the sensor output signal. The gas concentration of Z-3-hexen-1-ol was determined in the concentration range from 0.01 to 100 ppm with short response time of less than 1 s and a high reversibility of the sensor signal in ambient air.

The antennae of insects are covered with sensillae, among them the olfactory sensillae, which contain within them the olfactory receptor neurones (ORN). Many of the ORNs of insect species are interesting and challenging objects of investigation, because of their very low thresholds to ambient odourous chemicals.

Insect ORNs used for sensing odours were reported and the odour-response of individual ORN was recorded by microelectrodes (Huotari 2000). Action potentials of insect olfactory receptor neurons (ORN) were picked up with microelectrodes and amplified by a high-impedance amplifier connected to an oscilloscope, an audio amplifier, and an instrumentation amplifier. The amplified action potentials and odour exposure times were recorded on a DAT (digital audio tape) recorder for later analysis. This measurement of the stimulated action potential rates of the ORN of a blowfly (*Calliphora vicina*) made it possible to estimate odour concentrations to which the ORNs of the antenna were exposed. Odours of 1,4-diaminobutane (14DAB), 1-hexanol (HX), and butanoic acid (BA) were used in the experiments to stimulate various ORNs, which discharged action potentials. Spontaneous and stimulated action potential rates were analyzed by means of a time-to- voltage converter connected to a signal analyzer. The ORNs of the Calliphora antenna were sensitive to and specific for 14DAB, BA and HX, but not sensitive to the other odour chemicals tested. The lowest concentrations (about 1 ppb for 14DAB, 20 ppm for BA and 8 ppm for HX) caused the action potential rates to increase from the spontaneous level to distinct levels that were very suitable for both signal analysis and concentration measurement purposes.

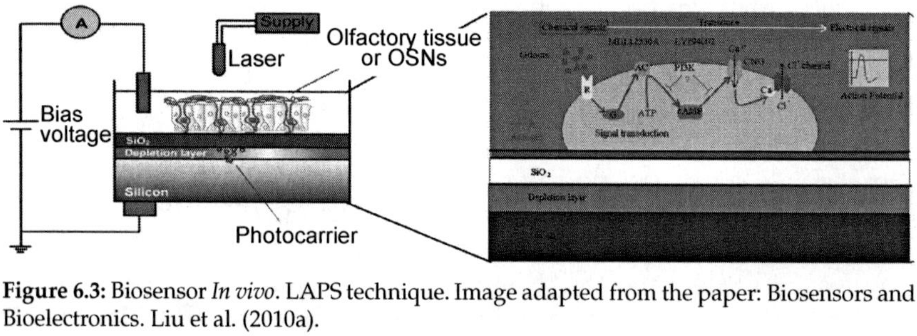

Figure 6.3: Biosensor *In vivo*. LAPS technique. Image adapted from the paper: Biosensors and Bioelectronics. Liu et al. (2010a).

Color image of this figure appears in the color plate section at the end of the book.

BIOELECTRONIC NOSES BASED ON OLFACTORY RECEPTORS

The olfactory biosensors have evolved successfully for the last few years. If we compare the biosensors that work with ORNs or cells transfected with ORs with those that work with intact olfactory epithelium we notice that several advantages exist such as simplicity, non-invasive detection and dispensing with sophisticated surgical operation (Ling et al. 2010).

Although the *in vivo* techniques, used to distinguish different odours, have greatly contributed to the studies of olfactory mechanism and ORN biosensors, they are merely elementary at present.

Because of their natural characteristics, ORs can be used to design a bioelectronic nose that mimics the olfactory system for the detection of odourants. ORs used as biological detectors are more sensitive in detecting their respective ligands than the most advanced physical approaches such as chemical noses or gas chromatography/mass spectrometry. Since ORs are G protein coupled receptors, olfactory receptors need to stay in their membrane environment to be functional.

There are two types of bioelectronic noses based on olfactory receptors: the cell-based and protein- based bioelectronic noses. The first type involves the use of ORs expressed in either their native olfactory sensory neurons (OSNs), such as those of rats and mice, or in heterologous systems such as yeast, amphibian, insect and mammalian cell lines.

As alternative ORs can be partially purified and used as recognition element in electronic noses.

The use of cells, cell extract or purified ORs as recognition element has a range of advantages over the use of the ORN biosensor *in vivo*. The most important ones are the level of miniaturization that the device can reach, the simplicity, controllability, noninvasive detection and dispensing with sophisticated surgical operation.

The use of a living cell in a biosensor device has the inconvenience of requiring specific conditions to maintain their health and function. Also, further reactions occurring in a cell and other elements might interfere with the signal being monitored.

Compared to the cultured olfactory cells, the intact olfactory epithelium can be obtained conveniently with the primary cell structure being well preserved.

Expression of OR for Biosensor Developments

Minic et al., have experience in olfactory receptor expression in yeast cells. Until now, it was known that S. cerevisiae yeasts are better than mammalian cells for olfactory receptor expression since they are much cheaper, easier to cultivate and the expression yield in yeast cells is higher than in mammalian cells. This strain has been successfully used for functional expression of many G-protein coupled receptors including the olfactory receptors (Minic et al. 2005a). The use of yeast cells for ORs expression, as eukaryotic organisms, have all the machinery necessary to allow the protein folding and trafficking.

The authors were involved in a framework of a European project called SPOT-NOSED in 2003. The aim was the development of a nanobiosensor based on single olfactory receptor, in order to mimic the natural olfactory system. For that reason they expressed two different receptors, ORI7 from a rat and OR1740 from a human in Saccahromyces cerevisiae yeast. The first problem they faced was how to improve and achieve the proper plasma membrane expression of the receptor because normally the ORs remain in the membrane of the endoplasmatic reticulum or were directed to the vacuoles to undergo proteolytic degradation. For that reason they made various adaptations of the classical way of GPCR expression on yeast cells (Minic et al. 2005b). In addition, it is reported that the tolerance of yeast cells to pH variations and temperature allows higher receptor functional expression.

The ORs expression procedure that they suggested improved the system using modified yeast strain, MC18 of *S. cerevisiae* with the luciferase gene. With this modification they could control the odourant stimulation by a sensitive bioassay, the functionality of this one and control where the olfactory receptor was mainly expressed by immuno analysis. Mini et al., also proved that working at 15°C during the yeast growth of a strain that was induced in galactose-containing medium allows an improvement of the functional response of the receptor. They developed a sensitive yeast-based bioassay for odourant screening where olfactory receptor in conjugation with a mammalian Gα subunit was functionally expressed in yeast cells.

Other authors also heterologously expressed other ORs such as *C. elegans* olfactory receptor ODR-10 but in HEK cells and applied other techniques such as Surface Plasmon Resonance technique to characterize the molecular interactions (Lee et al. 2006).

These days, the development of a specific analytical device for fast routine measurements in many files has been solicited for society. In addition, the use of micro- and nano-technologies has an important role in medicine. For this reason, the use of isolated olfactory receptors instead of whole-cells enables the researchers to develop an olfactory biosensor using the nano-technologies.

According to this proposal Vidic et al. (2006) found a way to work with nano-objects carrying olfactory receptors. They proposed that the yeast cells (carrying the ORs) could be mechanically disrupted and the purified membrane fraction obtained was isolated in the form of nanosomes with sizes between 40 nm to 700 nm. Additional sonication was performed on these spherical vesicles to homogenize them and reduce their sizes. What they obtained was a homogenous population of proteoliposomes with diameters approximately 100 nm called nanosomes. The maintenance of nanosomes homogeneity in the solution suggested that they can be used as a recognized part in a nanobioelectronic sensor device (Gomila 2006). They also proved the functionality of those olfactory receptors expressed in the nanosomes by Surface Plasmon Resonance (Vidic et al. 2006). Accordingly, it found a way to work with single ORs immobilized in a manner that they preserved their function. As ORs are hydrophobic as it was previously presented, a lipid or detergent environment to maintain their native conformation and function is required. This discovery was a revolution in the bioelectronic nose's field (Vidic 2010).

There are several groups in the world who are also focusing their studies on the development of a bioelectronic nose based on the electrical properties of olfactory receptors. They are also trying to optimize an efficient cell-free production of olfactory receptors that in the future could be of use in a bioelectronic nose platform. For more information see the following references: Kaiser et al. 2008, Lee and Park 2010.

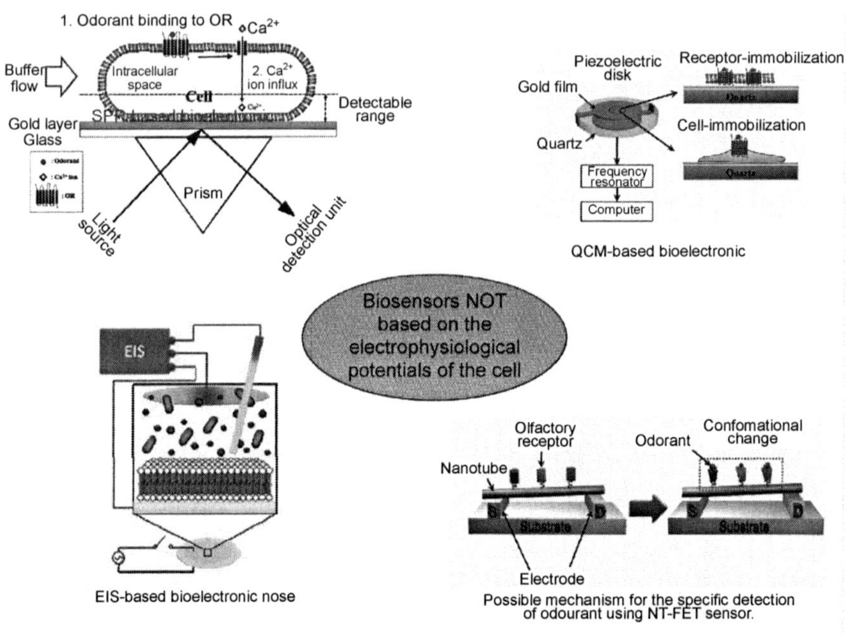

Figure 6.4: Biosensor. *In vitro*. Adapted from the papers: Biotechnol, Bioprocess Eng. Lee and Park (2010) and Lee et al. (2009a) Biosensors and Bioelectronics 25.

The Parameters Detected by the Sensors are Related to Changes in Intracellular Molecules Resulting from Signal Transduction

Cell-based biosensors, which use living cells or tissues as their sensitive elements, can produce signals that are related to biological sensation directly. In order to record the extracellular potentials of neurons, researchers developed microelectrode array (MEA) and light-addressable potentiometric sensor (LAPS) based on cell-based biosensors using microelectromechanical systems.

LAPS measure photocurrents generated when a site directed light source is applied (Owicki 1994). Changes in extracellular potential due to cell activity can generate modifications in the photocurrent signal that can be measured.

Studies have found that it was possible to detect the extracellular potential of receptor cells in intact epithelium surgically obtained from the rat, by keeping neuronal membrane in almost the same environment condition as *in vivo* (Ma et al. 1999, Kleene 2008).

The first example of an olfactory biosensor based on LAPS is from Liu et al. (2006). They investigated the extracellular potentials of primary cultured olfactory cells under odourant stimulation. This device was sensitive to environmental odour changes.

Years later the same group developed an olfactory neuronchip by combining the olfactory epithelium with LAPS (Liu et al. 2010a). The recording of the extracellular potentials of the olfactory receptor neurons in the epithelium was analyzed in response to odour stimulation. After stimulation by two different odourants, acetic acid and butanedione, olfactory epithelium activities were analyzed on the basis of local field potentials and presented different firing modes. The signals fired in different odourants could be distinguished into different clusters by principal component analysis (PCA).

Rat olfactory sensory neurons (OSNs) were cultured on the surface of LAPS chip and used as sensing elements (Wu et al. 2009). OSNs could respond to the stimulation of odours and convert chemical signals of odour molecules into electrical signals. To validate the response capacity of OSNs cultured on the LAPS chip to odours, an odour mixture was first used to stimulate OSNs. The response of OSNs to the mixture was recorded by the LAPS measurement system. The number of recorded firing spikes increased significantly. In order to confirm that the recorded firing spikes have originated from the olfactory signals of OSNs, MDL12330A, which is the specific inhibitor of adenyl cyclase, was used to inhibit the olfactory transduction pathway in OSNs and subsequently inhibits the responses of OSNs to the stimulations. The number of firing spikes decreased dramatically, which implies the responses of OSN to the mixture are inhibited by MDL12330A. The enhanced effect of LY294002 to the response of OSNs was also investigated, which is the specific inhibitor of phosphatidylinositol 3-kinase (PI3K). The results confirmed that the response of OSNs to the mixture was enhanced by LY2.

From techniques such as electro-olfactogram (EOG), based on the recording of an extracellular field potential produced by the odour-induced generator currents in the olfactory receptor neurons, on the whole animal and tissue, to using methods such as voltage clamping on single cells or patch clamping to monitor single channels (Sakmann and Neher 1984), the electrophysiological approach to these techniques forms the basis for a range of biosensor transduction mechanisms. Researchers have used EOG and EAG to act as a sensor transducer component. EAG for example was used on a female B-mori antenna to assess odorous contaminants in a laboratory (Ziesmann et al. 2000).

Due to the technical expertize required for some of these techniques, in addition to the need for miniaturization, stability and portability in a biosensor device, transducer systems utilizing planar microelectrodes (microelectrodes arrays) to monitor electrophysiological changes in the active cells are becoming more popular.

Microfabricated planar electrodes can be used to measure the change in the membrane potential of the OSNs generated after the interaction between odourant and receptor.

In contrast to LAPS, which are a single channel recording system, MEA can record the multisite potentials simultaneously. That means that MEA can detect signals of many cells synchronously, which is convenient to comparatively analyze recorded information in parallel.

Whole yeast cells expressing human receptor OR-1740 were immobilized onto an interdigitated thin film microelectrode and odour-specific changes in conductance were measured (Marrakchi et al. 2007). When yeast cells attached to the gold microelectrode surface, pre-treated with poly-lysine solution, were stimulated with the receptor ligand helional, it was possible to detect conductiometric changes due to the ionic exchanges resulting from the recognition of the ligand molecules by the olfactory receptor.

S. cerevisiae yeasts are more convenient than mammalian cells for olfactory receptor expression since they are much cheaper and easier to cultivate.

A microfabricated planar electrode connected to an oscilloscope was also used to measure the electrical responses generated after the interaction of octanal and ORI7 expressed in HEK-293 cells (Lee et al. 2009b). As a way of amplifying the signal for the biosensor, the HEK cells were produced coexpressing ORI7 with a cyclic nucleotide gated (CNG) channel, providing the opportunity for increased calcium influx from outside the cell upon octanal binding, therefore resulting in a larger change in membrane potential.

In an experiment conducetd by Liu et al. (2010b) 36-channel MEAs were fabricated on a glass substrate and olfactory epithelium was stripped from rats and fixed on the surface of MEAs. Electrophysiological activities of olfactory receptor neurons in intact epithelium were measured through the multi-channel recording system. The extracellular potentials of cell networks could be effectively analyzed by correlation analysis between different channels. After being stimulated by odourants, such as acetic acid and butanedione, the olfactory cells generate different firing modes. These firing characteristics can be derived by time-domain and frequency-domain analysis, and they were different from spontaneous potentials.

In a different work rat ORNs were cultivated on the surface of a 60-channel MEA device and in order to study the odour-response characteristics of ORNs, two types of odourants at different concentrations were quantitatively pumped into the neurochip by a customized gas intake system (Ling et al. 2010). The odour-response features of ORNs such as firing amplitude, firing threshold, firing rate as well as firing channels were analyzed quantitatively and qualitatively in terms of ORN spike trains. After introducing the classification algorithm based on the spike threshold, the odour-response maps from the multitude sites could be used to identify DL-limonene and isoamyle acetate odourants.

The Parameters Detected are not the Electrophysiological Potentials of the Cells

Odourant detection can be measured recording the interaction between the OR and the odourant molecules as a change in weight, light, sound, heat, chemical composition or electrochemical signal. These measurements can be made using a variety of techniques including surface plasmon resonance (SPR), quartz crystal microbalance (QCM) electrochemical impedance spectroscopy (EIS) and field-effect transistor (FET).

Detection by QCM

The first bioelectronic nose was based on the direct immobilization of olfactory cilia onto a sensor solid surface. In 1999 Wu immobilized a crude bullfrog cilia preparation onto a piezoelectric (PZ) electrode which served as a signal transducer (Wu 1999). Using this biosensor he was able to detect trace levels of various odourants with concentrations fully correlated with olfactory threshold values of human noses. The results indicate rapid (about 400s), reversible and long-term (up to 3 mon) stable response to different volatile compounds. Woo also fractionated olfactory receptors from the cilia preparation and coated them separately onto the PZ crystal surface. A quartz crystal microbalance was used for detecting the binding of odourant molecules to the ORs. QCM is a bulk acoustic wave sensor which by applying an alternating electric field induces a resonance in a quartz crystal. The crystal oscillates at a tuned frequency, which changes in accordance with the mass on the crystal. When a material is adsorbed on the crystal surface, the frequency of oscillation changes and this change can be measured. QCM can detect the interaction between various odourants and chemicals coated on the crystal surface. An array made of six sensors consisting of five different cilia fractions was able to rapidly and stably detect responses to different volatile compounds.

A study was conducted that demonstrated the high specificity and sensitivity of QCM to detect odourants (Ko and Park 2005). In this case ORI7 was expressed in HEK-293 cells which were then adsorbed on the gold surface of QCM. The ORs were exposed to various odourant vapours but only octanal, the specific odourant of ORI7, gave a high signal. This signal increased with the concentration of the odourant.

Sung et al. coated QCM crystal surface with crude membrane extracts of *E.coli* expressing the receptor protein ODR-10 and examined its interactions with various odourant molecules (Sung et al. 2006). They showed linear dose-dependent responses upon stimulation of the membrane extracts with the natural receptor ligand diacetyl. This work demonstrated how membrane extracts, containing ORs, can retain full activity and can discriminate its specific odourant.

Detection by SPR

SPR spectroscopy is a technique based on the measurement of small changes of the resonance angle (θ_{SPR}) between incident light and the plasmon present on a thin metal layer. This value depends on the refractive index (RI) at the solid-liquid interface. Hence the SPR angle shifts are dependent upon the amount of material adsorbed on the gold surface, generating the corresponding analytical signal. SPR technique has been used to characterize biomoleclar interactions without labelling. SPR is a label-free and high sensitive optical sensing method. However, despite its advantages, SPR selectivity is rather low and it does not provide a cheap and portable transducer solution.

The binding between ORs and odourants has been examined by SPR by many research groups.

ORI7 and OR1740 were co-expressed with $G_{\alpha olf}$ subunit in *S. cerevisiae* by Vidic et al. (2006). Nanometer size vesicles, called nanosomes, were obtained and immobilized on a BIacore sensor chips L1. Nanosomes were grafted thanks to the presence of lipophilic alkyl chains in the dextran polymer that covers the L1 chip. The binding of

OR-odourants was followed by SPR by measuring the activation of the trimeric G protein and the consequent mobilization of the $G_{\alpha olf}$ subunit in presence of GTP. When the specific odourant together with GTP was injected onto the nanosomes, a significant SPR signal was detected. On the other hand, when the odourant and GTP were alone and a non-specific odourant together with GTP were used, no SPR signal was observed as a consequence of a missed activation of the ORs.

Moreover the sensitivity of the experiment was enhanced by replacing the GTP with its non-hydrolysable analogue, $GTP_{\gamma}S$.

This study showed that ORs can be expressed in yeast cells and retain their full activity in isolated nanosomes immobilized on sensorchips. The specific odourant was discriminated among odourants with a similar structure and different concentrations of the odourant could be measured.

The SPR technique was also used in the measurement of odourant molecules in a cell-based biosensing technique (Lee et al. 2009a). ORI7 was fused with a rho-tag import sequence at the N-terminus of ORI7 and expressed on the surface of human embryonic kidney (HEK)-293 cells. The cells were then immobilized on a SPR sensor chip treated previously with poly-D-lysine. Octanal, the specific ORI7 odourant, was then injected. The SPR signal was linearly dependent on the amount of injected odourant. Since the distance between the OR expressed on the cell surface and the gold surface is longer than the SPR detectable range, the response cannot be associated to the simple binding between the odourant and the OR. The binding event must trigger intracellular signal transduction, which causes changes in intracellular components. These intracellular changes may generate a change in SPR signal. In fact only when the odourant was used together with a Ca^{2+} solution was the SPR response observed. This means that the binding of odourant with OR induces Ca^{+} influx through intrinsic transmembrane ion channels and this influx of Ca^{+} ions into cells generated the SPR signal change.

Detection by EIS

Electrochemical impedance spectroscopy is a technique that measures the electrical resistance of a system and changes which occur in this resistance due to changes at a transducer surface (Lisdat and Schäfer 2008).

In a study by Hou et al. (2007) ORI7 were expressed in yeast and its membrane fractions were immobilized on a gold electrode. The immobilization technique made use of a mixed self-assembled monolayer and of a biotinylated ORI7-specific antibody. The odourant detection was electronically measured in a quantitative manner by electrochemical impedance spectroscopy measurements on samples (octanal and heptanal) and controls (helional). The response for the specific detection of odourants are higher than those obtained in the negative control tests, giving a higher variation in impedance. The corresponding biosensor was stable during 7 d. This study demonstrated that the electrical properties of the receptor in its lipidic environment can be modified by the process of odourant recognition and that these modifications can be detected by EIS measurements.

Another example of electrochemical study of ORs stimulation by odourants, is represented by a study from Benilova et al. (2008). OR1740 was expressed in yeast and the membrane fractions were anchored to a gold substrate modified with thiol-based multilayer. The ORs were specifically immobilized via a biotinylated antibody specific for the cmyc tag carried by the ORs. Under injections of helional (specific odourant of OR1740) activation of the OR was electrochemical revealed. Applying GTP-γ-S at 4°C as a specific enhancer of biochemical signal, it became possible to detect 10^{-11}M helional. The low temperature improved the lifetime of the biosensor by 2 d of continuous work. The selectivity was also tested by using the non-specific odourant heptanal which did not give any shift of the impedance spectrum.

Detection by FET

FETs have the ability of measuring ion concentrations and cell potential changes. Single walled nanotubes represent attractive components for use in FET devices due to their size and electrical properties. They provide great potential to produce low-cost, large-area, lightweight and flexible devices. Carbon nanotubes can be used to immobilize ORs onto a device. When the odourant binds the OR, a charge transfer occurs in the OR and the response is monitored by the change in the current of nanotubes.

The first example of FET-type bioelectronics nose based on human olfactory receptors and nanotubes (hOR-conjugated conducting polymer) is represented by a study from Yoon et al. in which they express hOR2AG1 in *E. coli* with a glutathione-S-transferase (GST) tag at its N terminus (Yoon et al. 2009). The *E. coli* cells were sonicated to obtain membrane fractions. Carboxylated polypyrrole nantotubes (CPNT) were synthesized and immobilized on an interdigitated microelectrode array (IDA) of gold microelectrodes generating a FET device. The IDA substrate was treated with an aminosilane and the carboxy functional groups of the CPNTs were then coupled with the surface amino group of the IDA. hOR2AG1 were conjugated to the CPNTs through the covalent interaction between the terminal amine groups on cysteine residues of OR and the carboxy groups of CPNTs. hOR2AG1 specifically responds to amyl butyrate, a common reagent for fruit flavour. The FET showed a concentration-dependent increase in current change (I_{SD}) when exposed to amyl butyrate. It is believed that the current increase is due to a change in the charge-transport behaviour of the CPNTs. Specifically, the cysteine residues of hORs adopt an uncharged (RSH) state or negative charged (RS^-) state, associated with the acid-to-base transition of the sulphydryl group (Neuhaus et al. 2006). The specific binding of odourants initiates the structural rearrangement of hORs finally leading to an increase in negatively charged base state (RS^-). The increased negative charge density at the OR-CPNT interface can induce the increase in the I_{SD}. When other odourants different from amyl butyrate were used, no significant changes in I_{SD} were observed showing the specificity of this FET-type bioelectronic nose. The detection limit of the FET was found to be 10 fM, which is two orders of magnitude lower than those of other bioelectronics noses (Sung et al. 2006). Single-walled carbon nanotubes were also used to develop a FET-olfactory sensor system (Kin et al. 2009).

hOR2AG1 was expressed in *E.coli* and coated on swCNT-FET and the deformation of the hOR2AG1 protein upon binding specific odourant molecules was detected by the swCNT-FET. The detection limit was 100fM and the high selectivity was tested by molecules with a chemical structure very similar to that of amyl butyrate.

CONCLUSIONS

The use of smell in different fields has been rediscovered due to major advances in odour sensing technology and artificial intelligence. However, current electronic noses, based on electronic sensors, have significant limitations concerning sensitivity, reliability and selectivity, amongst others. These limitations are the basis of recurrent problems of this technology to reach essential applications in different areas, with a high potential in the diagnosis of different diseases.

Mass spectroscopy analysis and the use of dogs to distinguish between samples of patients and samples control confirm that this approach can be useful for medical diagnostic purposes if the electronic nose development allows distinguishing the odourant patterns with high accuracy and short measurement time. New bioelectronic noses based on olfactory receptors in order to mimic the animal nose could be an interesting approach combining nano, bio and information technologies.

REFERENCES

Angioy, A.M. and A. Desogus, I.T. Barbarossa, P. Anderson and B.S. Handsson. 2003. Extreme sensitivity in an olfactory system. Chem Senses 284, 279.

Baker, T.C. and K.F. Haynes. 1989. Field and laboratory electroantennographic measurements of pheromone plume structure correlated with oriental fruit moth behaviour. Physiol Entomol 14: 1–12.

Benilova, I. and V. Vidic, J. Pajot-Augy, E. Soldatkin, A.P. Martelet, C. Jaffrezic-Rneaults and N. Mater. 2008. Electrochemical study of human olfactory receptor OR 17-40 stimulation by odorants in solution. Sci Eng C 28: 633.

Breer, H. 2003. Olfactory receptors: molecular basis for recognition and discrimination of odours. Anal Bioanal Chem 377: 427–433.

Buck, L. and R. Axel. 1991. A novel multigene family may encode odorant receptors: a molecular basis for odour recognition. Cell 65: 175–187.

Chen, X. et al. 2005. A study of an electronic nose for detection of lung cancer based on a virtual SAW gas sensors array and imaging recognition method. Meas Sci Technol 16, 1535.

Chesler, A. and S. Firestein. 2008. Current views on odour receptors. Nature 452, 944.

Church, J. and H. Wiliams 2001. Another sniffer dog for the clinic? Lancet 358, 930.

Cornu, J.N., Cancel-Tassin et al. 2011. Olfactory detection of prostate cancer by dogs sniffing urine: a step forward in early diagnosis Eur Urol doi :10.1016/j.eururo.2010.10.10.006

Dalton, P. and A. Gelperin and G. Preti. 2004. Volatile metabolic monitoring of glycemic status in diabetes using electronic olfaction. Diabetes Technol Ther 6, 534.

DeMaria, S. and J. Ngai. 2010. The cell biology of smell. The Journal of Cell Biology Vol. 191 n°3, 443–452.

Di Natale, C. et al. 1999. Electronic nose analysis of urine samples containing blood. Physiol Measurement 20: 377–384.

Di Natale, C. et al. 2003. Lung cancer identification by analysis of breath by means of an array of non-selective gas sensors. Biosensors and Bioelectronics 18: 1209–1218.

Dulac, C. and R. Axel. 1995. A novel family of genes encoding putative pheromones receptors in mammals. Cell 83: 195–206.

Elliker, K.R. and B.A. Sommerville and D.M. Broom. 2004. Dogs as doctors: training dogs to detect prostate cancer. Paper presented at the Chemical Signals in Vertebrates Workshop XII, Withersdane Hall, Wye, England.

Davide, F. and M. Holmberg and I. Lindström. Communications Through Virtual Technology. p. 194. *In:* G. Riva and F. Davide. (eds). 2001. Identity Community and Technology in the Internet Age. IOS Press: Amsterdam, The Netherlands.

Fiñipiak, W. and A. Sponring and A. Filipiak et al. 2010. TD-GC-MS analysis of volatile metabolites of human cancer and normal cells *in vitro*. Cancer Epidemiol Biomarkers Prev 19: 182–195.

Firestein, S. 2001. How the olfactory system makes sense of scents. Nature 413: 211–217.

Gardner, J.W. and P.N. Bartlett. 1999. Electronic Noses: Principles and Applications. Oxford Univ Press, UK.

Glatz, R. and K. Bailey-Hill. 2011. Mimicking nature's noses: From receptor deorphaning to olfactory biosensing. Progress in Neurobiology 93: 270–296.

Gomila, G. et al. 2006. Advances in the production, immobilization, and electrical characterization of olfactory receptors for olfactory nanobiosensor development. Sensors and Actuators B. Chemical. Vol. 116 n°1–2, 66–71.

Hanson C.W. and H.A. Steinberger. 1997. The use of a novel electronic nose to determine the etiology of intrapulmonary infection [abstract]. Anesthesiology 87 A, 269.

Hengl, T. et al. 2010. Molecular components of signal amplification in olfactory sensory cilia. PNAS 107 n° 13: 6052–6057.

Horvath, and et al. 2008. Human ovarian carcinomas detected by specific odor Integr. Cancer Ther 7: 76–80.

Hou, Y. et al. 2007. A novel detection strategy for odorant molecules based on controlled bioengineering of rat olfactory receptor I7 Biosens Bioelectron 22: 1550, 2007

Huotari, M.J. 2000. Biosensing by insect olfactory receptor neurons. Sens. Actuators B: Chem 71, 212.

Kaiser, L. et al. 2008. Efficient cell-free production of olfactory receptors: Detergent optimization, structure, and ligand binding analyses. PNAS 105 n° 41: 15726–15731.

Kaupp, U.B. 2010. Olfactory signalling in vertebrates and insects: differences and commonalities. Nature 11: 188–200.

Kin, T.H. et al. 2009. Single-Carbon-Atomic-Resolution Detection of Odorant Molecules using a Human Olfactory Receptor-based Bioelectronic Nose Adv Mater 21: 91, 2009.

Kleene, S.J. 2008. The Electrochemical Basis of Odor Transduction in Vertebrate Olfactory Cilia Chem Senses 3: 839–859.

Ko, H.J. and T.H. Park. 2005. Piezoelectric olfactory biosensor: ligand specificity and dose-dependence of an olfactory receptor expressed in a heterologous cell system. Biosens Bioelectron 20: 1327, 2005.

Ko, H.J. and T.H. Park. 2008. Enhancement of odorant detection sensitivity by the expression of odorant-binding protein. Biosens Bioelectron 23: 1017–1023.

Ko, H.J. et al. 2010. Specificity of odorant-binding proteins: a factor influencing the sensitivity of olfactory receptor-based biosensors. Bioprocess Biosyst Eng 33: 55–62.

Lacey, R.E. and G.S. Osborn. 1998. Application of Electronic Noses in Measuring Biological Systems. 1998 ASAE Annual International Meeting Sponsored by ASAE Disney's Coronado Springs Resort Orlando, Florida.

Lee, J.Y. et al. 2006. Cell-based measurement of odorant molecules using surface Plasmon resonance. Enzyme and Microbial Technology 39 n°3 475–380 Biotechnology and Bioprocess Engineering 15:1 22–29.

Lee, S.H. and H.J. Ko and T.H. Park. 2009a. Real-time monitoring of odorant-induced cellular reactions using surface plasmon resonance Biosens Bioelectron 25, 55.

Lee, S.H. and S.B. Jun H.J. Ko, S.J. Kim and T.H. Park. 2009b. Cell-based olfactory biosensor using microfabricated planar electrode. Biosens Bioelectron 24, 2659.

Lee, S.H. and T.H. Park. 2010. Recent Advances in the Development of Bioelectronic Nose. Biotechnology and Bioprocess Engineering 15: 22–29.

Liberles, S.D. and L. Buck. 2006. A second class of chemosensory receptors in the olfactory epithelium. Nature 442: 645–650.

Ling, S. et al. 2010. The fabrication of an olfactory receptor neuron chip based on planar multi-electrode array and its odor-response analysis. Biosensors and Bioelectronics 26: 1124–1128.

Lisdat, F. and D. Schäfer. 2008. The use of electrochemical impedance spectroscopy for biosensing. Anal Bioanal Chem 391: 1555–1567.

Liu, Q. and H. Cai, Y. Xu, Y. Li, R. Li and P. Wang. 2006. Olfactory cell-based biosensor: A first step towards a neurochip of bioelectronic nose. Biosens Bioelectron 22, 318.

Liu, Q. and W. Ye, N. Hu, H. Cai, H. Yu and P. Wang. 2010a. Olfactory receptor cells respond to odors in a tissue and semiconductor hybrid neuron chip. Biosensors and Bioelectronics 26: 1672–1678.

Liu, Q., W. Ye, L. Xiao, L. Du, N. Hu, P. Wang. 2010b. Extracellular potentials recording in intact olfactory epithelium by microelectrode array for a bioelectronic nose. Biosens Bioelectron 25, 2212.

Ma, M. and W.R. Chen and G.M. Shepherd. 1999. Electrophysiological characterization of rat and mouse olfactory receptor neurons from an intact epithelial preparation J Neurosci Methods 92: 31–40.

Machado, R.F. et al. 2005. Detection of Lung Cancer by Sensor Array Analyses of Exhaled Breath. Am J Respir Crit Care Med 171, 1286.

Malnic, B. et al. 1999. Combinatorial receptor codes for odors. Cell 965, 713.

Marrakchi, M. and J. Vidic, N. Jaffrezic-Renault, C. Martelet and E. Pajot-Augy 2007. A new concept of olfactory biosensor based on interdigitated microelectrodes and immobilized yeasts expressing the human receptor OR17–40 Eur Biophys J 36, 1015.

Matsunami, H. and A. Buck. 1997. A multigene family encoding a diverse array of putative pheromone receptors in mammals. Cell 90: 775–784.

McCulloch, M., T. Jezierski, M. Broffman et al. 2006. Diagnostic accuracy of canine scent detection in early and late stage lung and breast cancers. Integr Cancer Ther 5: 30–39.

Minic, J. et al. 2005a. Yeast systems as a screening tool for pharmacological assessment of G-protein coupled receptors. Current Medical Chemistry 12 n°.8, 961–969.

Minic, J. 2005b. Functional expression of olfactory receptors in yeast and development of a bioassay for odorant screening. The FEBS Journal 272: 524–537.

Mombaerts, P. and M. Nei. 2004. Genes and ligands for odorant, vomeronasal and taste receptors. Nature Rev Neurosci 5: 263–278.

Neuhaus, E.M. and A. Mashukova, W. Zhang, J. Barbour and H. Hatt. 2006. A Specific Heat Shock Protein Enhances the Expression of Mammalian Olfactory Receptor Proteins. Chem Senses 31: 445–452.

Ortega, A. and S. Marco and T. Sundic et al. 2000. New pattern recognition systems designed for electronic noses. Sensors and Actuators B-Chemical 69: 302–307.

Owicki, J.C. and L.J. Bousse, D.G. Hafeman, G.L. Kirk, J.D. Olson, H.G. Wada and J.W. Parce. 1994. The Light-Addressable Potentiometric Sensor: Principles and Biological Applications. Ann Rev Biophys Biomol Struct 23: 87–113.

Pavlou, A. et al. 2000. An *in vitro* rapid odour detection and recognition model in discrimination of *H. pylori* and other gastroeosophageal pathogens. Biosensors and Bioelectronics 15: 333–342.

Pavlou, A. et al. 2002. Use of an electronic nose system for diagnoses of urinary tract infections *in vivo*. Biosensors and Bioelectronics 17: 893–899.

Pearce, T.C. and S.S. Schiffman, H.T. Nagle and J.W. Gardner (eds.). 2002. Handbook of Machine Olfaction: Electronic Nose Technology. Weinheim, Germany: Wiley-VCH, 2002.

Pelosi, P. 1994. Odorant- Binding Proteins. Critical Reviews in Biochemistry and Molecular Biology 29(3): 199–228.

Pelosi, P. et al. 2005. Diversity of Odorant-binding Proteins and Chemosensory Proteins in Insects Chem Senses 30: 291–292.

Peng, G., M. Hakim, Y.Y. Broza et al. 2010. Detection of lung, breast, colorectal and prostate cancers from exhaled breath using a single array of nanosensors. Br J Cancer 103: 542–551.

Persaud, K. and G. Dodd. 1982. Analysis of discrimination mechanisms in the mammalian olfactory system using a model nose. Nature 299: 352–355.

Riviere, S. et al. 2009. Formyl peptide receptor-like proteins are a novel family of vomeronasal chemosensors. Nature 459: 574–577.

Sakmann, B. and E. Neher. 1984. Patch Clamp Techniques for Studying Ionic Channels in Excitable Membranes. Annu Rev Physiol 46: 455–472.

Sato, K. and et al. 2008. Insect olfactory receptors are heteromeric ligand-gated ion channels. Nature 452: 1002–1006.

Sauer, A.E. and G. Karg, U.T. Koch, J.J. De Kramer and R. Milli. A portable EAG system for the measurement of pheromone concentrations in the field. 1992. Chem Sens 17: 543–553.

Schöning, M.J. and S. Schütz P. Schroth, B. Weißbecker, A. Steffen, P. Kordo, H.E. Hummel and H. Lüth. 1998. A BioFET on the basis of intact insect antennae. Sens. Actuators B: Chem 5, Shoning, M.J. 47, 235.

Schütz, S. and B. Weiβbecker and H.E. Hummel. 1996. Biosensor for volatiles released by damaged plants. Biosens Bioelectron 11: 427–433.

Song, H.G. et al. 2008. First contact to odours: our current knowledge about odorant receptors. Sensors 8: 6303–6320.

Sonoda, H. and S. Kohnoe and T. Yamazato, et al. 2011. Colorectal cancer screening with odour material by canine scent detection. Gut. Doi: 10.1136/gut.2010.218305

Spanel, P. et al. 1999. Analysis of Formaldehyde in the Headspace of Urine from Bladder and Prostate Cancer Patients Using Selected Ion Flow Tube Mass Spectrometry. Rapid Commun Mass Spectrom 13, 1354.

Spehr, M. and D. Munger. 2009. Olfactory receptors: G protein-coupled receptors and beyond. Journal of Neurochemistry 109: 1570–1583.

Su, C.Y. et al. 2009. Olfactory Perception: Receptors, Cells, and Circuits. Cell 139: 45–59.

Sung, J.H. and H.J. Ko and T.H. Park. 2006. Piezoelectric biosensor using olfactory receptor protein expressed in *Escherichia coli.* Biosens Bioelectron 21.

Tegoni, M. et al. 2000. Mammalian odorant binding proteins. Biochimica et Biophysica Acta 1482: 229–240.

Touhara, K. and L.B. Vosshall. 2009. Sensing odorants and pheromones with chemosensory receptors. Annu Rev Physiol 71: 307–332.

Turner, A.P.F. 2000. Biosensors , sense and sensitivity. Science 290: 1315–1317.

Turner, A.P.F. and N. Magan. 2004. Electronic noses and disease diagnostics. Nature Reviews/Microbiology 2: 161–166.

Vidic, J. 2010. Bioelectronic Noses Based on Olfactory Receptors. pp. 377–386 *In:* Intelligent and Biosensors, Croatia.

Vidic, J. et al. 2006. Quantitative assessment of olfactory receptors activity in immobilized nanosomes: a novel concept for bioelectronic nose. Lab on a Chip 6: 1026–1032.

Vidic, J. et al. 2008. On a chip demonstration of a functional role for Odorant Binding Protein in the preservation of olfactory receptor activity at high odorant concentration. Lab Chip 8(5): 678–688.

Vogt, R. and L. Riddiford. 1981. Pheromone binding and inactivation by moth antennae. Nature 293: 161–163.

Vosshall, L.B. and M.C. Stensmyr. 2005. Wake Up and Smell the Pheromones Neuron 45: 179–187.

Wang, G.R. et al. 2003. Cloning, expression and immunocytochemical localization of a general odorant-binding protein gene from Helicoverpa armigera (Hubner). Insect Biochem Mol Biol 33: 115–124.

Wicher, D. et al. 2008.Drosophila odorant receptors are both ligand-gated and cyclic-nucleotide-activated cation channels. Nature 452: 1007–1011.

Williams, H. and A. Pembroke. 1989. Sniffer dogs in the melanoma clinic? Lancet 1734.

Willis, C.M., S.M. Church, C.M. Guest et al. 2004. Olfaction detection of human bladder cancer by dogs: proof of principle study BMJ 329, 712.

Wilson, A.D. and M. Baietto. 2011. Advances in Electronic-Nose Technologies Developed for Biomedical Applications. SENSORS 11, 1: 1105–1176.

Wilson, A.D. and M. Baietto. 2009. Applications and Advances in Electronic-Nose Technologies SENSORS: 9: 5099–5148.

Wu, C. and P. Chen, H. Yu, Q. Liu, X. Zong, H. Cai and P. Wang. 2009. A novel biomimetic olfactory-based biosensor for single olfactory sensory neuron monitoring. Biosens. Bioelectron 24, 1498.

Wu, T. 1999. A piezoelectric biosensor as an olfactory receptor for odour detection: electronic nose. Biosensors & Bioelectronics 1: 14: 9–18.

Wyatt, T.D. 2010. Pheromones and signature mixtures: defining species-wide signals and variable cue for identity in both invertebrates and vertebrates. J Comp Physiol A 196: 685–700.

Yoon, H. et al. 2009. Polypyrrole Nanotubes Conjugated with Human Olfactory Receptors: High-Performance Transducers for FET-Type Bioelectronic Noses. Angewandte Chemie Int edn 48, 2755.

Zarzo, M. 2007. The sense of smell: molecular basis of odorant recognition. Biological Reviews 82: 455–479.

Zhou, J.J. 2010. Odorant-Binding Proteins in insects. Vitamins and Hormones, Vol. 83: 241–272.

Zhuang, H. and H. Matsunami. 2008. Evaluating cell-surface expression and measuring activation of mammalian odorant receptors in heterologous cells. Nature Protocols Bol 3 n°.9 1402–1413.

Ziesmann, J. and I. Valterova, K. Haberkorn, M.G. de Brito Sanchez and L-E. Kaissling. 2000. Chemicals in Laboratory Room Air Stimulate Olfactory Neurons of Female *Bombyx mori*. Chem Senses 25: 31–37.

7

Nanomedicine Technologies for Cell-based Drug Screening

Pravin K. Naoghare[1,2] and *Joon Myong Song*[1,a,*]

INTRODUCTION

Nanomedicine is the application of nanotechnology to medicine. The word 'nano' is taken from the Greek word nanos or dwarf which means one-billionth 10^{-9} of something. Nanomedicine utilizes nano-sized materials/tools for the diagnosis, prevention and treatment of disease so as to gain better knowledge of the complex pathophysiology of disease (Freitas 2005). Nanomedicine can also be defined as the comprehensive monitoring, control, construction, repair, defense and improvement of all human biological systems at the molecular level using engineered nanodevices and nanostructures (Morrow et al. 2007). Nanomedicine exploits nano-tools to provide speed and high performance for elucidating the complex biological phenomenon. Nanomedicine technologies include dendrimers, carbon fullerenes (buckyballs), nanoshells, quantum dots etc., to target specific bio-molecules (Patil et al. 2008). The most advanced form of nanomedicine technologies are nanorobots and nanoinstruments. These nanomedicine technologies are still at the developmental stage.

[1]Research Institute of Pharmaceutical Sciences and College of Pharmacy, Seoul National University, Seoul 151-742, South Korea.

[a]E-mail: jmsong@snu.ac.kr

[2]Current Address: Environmental Health Division, National Environmental Engineering Research Institute, Nehru Marg, Nagpur-20, India.

*Corresponding author

With the help of these machines we will be able to repair or replace the damaged cells. The quantum dot is one of the promising nanomedicine technologies which have been widely used for biological applications. Quantum dots are particularly significant for optical applications due to their high extinction co-efficient and photostability. Figures 7.1 and 7.2 illustrate the structure and the size of the quantum dot, respectively.

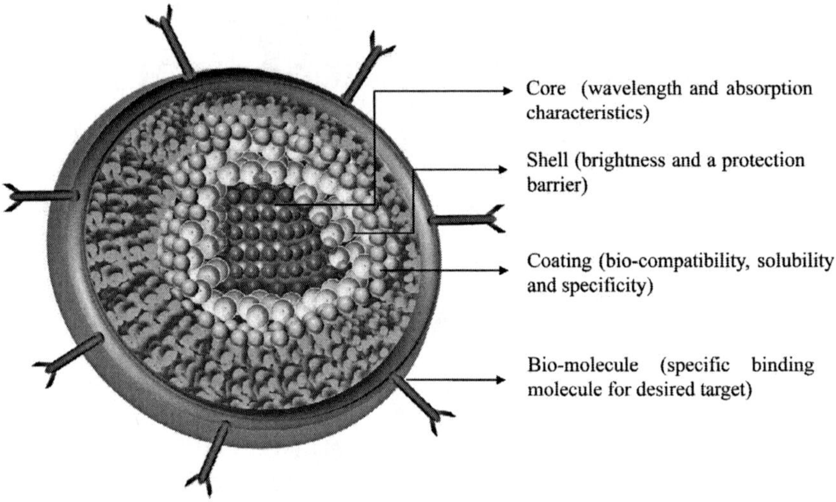

Figure 7.1: Structure of quantum dot.

Color image of this figure appears in the color plate section at the end of the book.

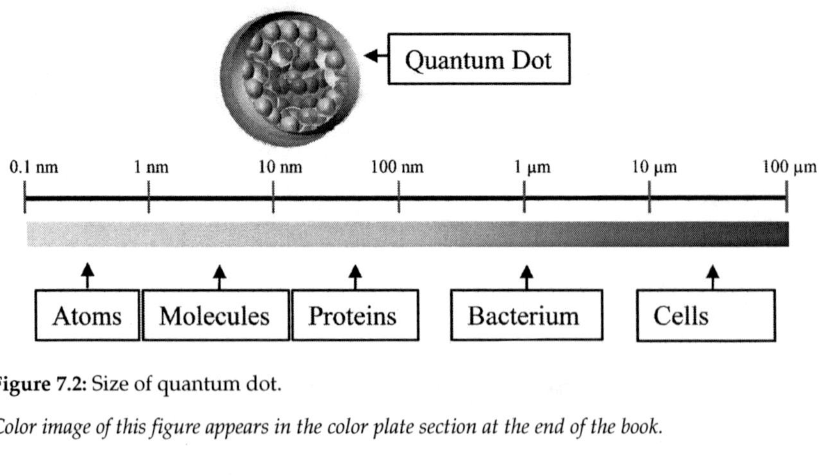

Figure 7.2: Size of quantum dot.

Color image of this figure appears in the color plate section at the end of the book.

NANOMEDICINE, WHY IS IT DIFFERENT?

Conventional medicine approaches differ from the new approaches in terms of the exploitation of properties of materials at nanoscale. The conventional approach mostly deals with the problems at tissue level based on the symptoms of the disease. This approach has potential drawbacks because the time when the symptoms become visible, the disease may already be at an advanced stage. In contrast, nanomedicine aims to detect the alterations due to biological dysfunction at molecular and cellular levels so as to provide early diagnosis to increase the survival rates and improve prognosis of the disease stage. For example, delivery of nanomedicine in to the targeted cell enables the efficient and accurate action of drugs with fewer side effects. On the onset, nanomedicine technology was used for diagnostic and therapeutic purposes. As this technology matured, complex nanodevices and even nanorobots were fabricated (He et al. 2010).

WHAT ARE NANOMEDICINE TECHNOLOGIES?

Nanomedicine technologies could find an enhanced position in various areas and applications of the healthcare sector including drug delivery, drug discovery screening and development, diagnostics and medical devices. Table 7.1 illustrates nanomedicine technologies that are either in the developmental stage or in the market.

NANOMEDICINE ADVANTAGES AND APPLICATIONS

The above-mentioned nanomedicine technologies have a great significance in different fields of research for the betterment of healthcare. These includes the application of nanomedicine technologies in diagnosis and treatment of diseases (Wee et al. 2005, Zheng et al. 2005), nervous system tracking, drugs dispersion, drug discovery, heart and ECG machine, artificial antibodies, surgery, biomolecular imaging, BioMEMS, proteomics and DNA-chips, biosensors (Zheng et al. 2005), gene therapy, stem cells, cloning etc. With these technologies conventional procedures can be transformed to achieve high-throughput and accuracy. Some of the examples of the application of nanomedicine technologies for healthcare are given in Tables 7.2 and 7.3.

Table 7.1: Nanomedicine Technologies (Saha 2009).

Raw nanomaterials	**Cell simulations and cell diagnostics**	**Biological research**
Nanoparticle coatings	Cell chips	Nanobiology
Nanocrystalline materials	Cell stimulator	Nanoscience in life sciences
Nanostructured materials	DNA manipulation, sequencing	Drug delivery
Cyclic peptides	diagnostics	Drug discovery screening
Dendrimers	Genetic testing	Biopharmaceutics
Detoxification agents	DNA microarrays	Drug encapsulation
Drug encapsulation	Ultrafast DNA sequencing	Smart drugs
Fullerenes	DNA manipulation and control	Molecular medicine
Functional drug carriers Smart drugs	Tools and diagnostics	Genetic therapy and Pharmacogenomics
Nanobarcodes	Bacterial detection systems and Biochips	Artificial enzymes and enzyme control
Molecular medicine	Biomolecular imaging	Enzyme manipulation and control
Nanoemulsions	Biosensors and biodetection	Nanotherapeutics
Nanofibers	Diagnostic and defense applications	Antibacterial and antiviral nanoparticles
Nanoparticles	Endoscopic robots and microscopes	Fullerene-based pharmaceuticals
Nanoshells	Fullerene-based sensors	Photodynamic therapy
Carbon nanotubes	Imaging (cellular, etc.)	Radiopharmaceuticals
Noncarbon nanotubes	Monitoring	Synthetic biology and early nanodevices
Quantum dots	Lab on a chip	Dynamic nanoplatform nanosome
Artificial binding sites	Nanosensors	Tecto-dendrimers
Artificial antibodies	Point of care diagnostics	Artificial cells and liposomes
Artificial enzymes, receptors	Protein microarrays	Polymeric micelles and polymersomes
Molecularly imprinted polymers	Scanning probe microscopy	Biotechnology and biorobotics
Control of surfaces	Intracellular biocomputers	Biologic viral therapy
Artificial surfaces-adhesives	Intracellular sensors/reporters	Virus-based hybrids
Artificial surfaces—nonadhesive	Implants inside cells	Stem cells and cloning
Biocompatible surfaces	Implantable materials and devices	Tissue engineering
Biofilm suppression	Implanted bioMEMS, chips, and electrodes MEMS/ Nanomaterials-based prosthetics	Artificial organs
Engineered surfaces	Sensory aids (artificial retina, etc.)	Nanobiotechnology
Thin-film coatings	Microarrays	Biorobotics and biobots
Nanopores	Microcantilever-based sensors	Nanorobotics
Immunoisolation	Microfluidics	DNA-based devices and nanorobots
Molecular sieves and channels	Microneedles	Diamond-based nanorobots
Nanofiltration membranes	MEMS surgical devices	Cell repair devices

Table 7.2: Application of nanomedicine for the healthcare.

	Nanomaterial Name & Type	**Pharmacological function**	**Diseases**
Nanomedicines (in clinic)	Liposome (30–100 nm)	Targeted drug Delivery	Cancer
	Nano-particle (Iron oxide, 5–50 nm)	Contrast agent for magnetic resonance imaging	Hepatic (Liver)
Nanomedicines (under development)	Dendrimer (5–50 nm)	Contrast agent for magnetic resonance imaging	Cardiovascular Phase III clinical trial
	Fullerene (Carbon bucky ball, 2–20 nm)	Antioxidant	Neurodegenerative, Cardiovascular
	Nanoshells (Goldcoated Silica, 60 nm)	Hyperthermia	Cancer Preclinical

Nanomedicine technologies have also been used for the treatment of cancer. Various pharmaceutical companies all over the world are marketing nanonedicine technologies under different trademarks. Table 7.3 illustrates some of the nanomedicine technologies for the treatment of cancer that are either in a developmental stage or available in the market.

IMPORTANCE OF NANOMEDICINE IN CELL-BASED DRUG DISCOVERY SCREENING

Pharmaceutical companies often deal with a screening of large numbers of compounds with the help of biochemical assays in an ultrahigh-throughput format. However, the drug treatment may trigger complex biological pathways which cannot be predicted using biochemical assays. The complexities in biochemical assays have resulted in the increased use of cell-based screening assays to elucidate the drug response with the help of different cellular assays.

Cell-based assays facilitates an important tool in basic research applications and drug screening techniques for the elucidation of complex biological phenomenon for better understanding of drug-cell interactions (Lunn 2010). Cellular assays have been routinely used to validate previously unknown targets, monitor cellular pathways and screen drug analogs for better efficacy and bio-safety. Cellular assays are now extensively used to determine drug-mediated toxicity and cellular responses and to screen better drug candidates in the initial stage of drug discovery using *in vivo* and *in vitro* models. Recently, rapid *in vitro* screening methods were used

Table 7.3: Nanotechnology cancer treatments that are being tested or have been approved.

Product	Generic name/description	Type of nanomaterial	Indication	Status	Company
VivaGel®	Polylysine dendrimer gel for topical application	Dendrimer derivatised with sodium naphthalene disulfonate	vaginal microbicide to prevent sexually transmitted infections	Phase II	StarPharma, Australia.
MRX-952	Liposome-encapsulated 7-ethyl-10-hydroxycamptothecin	Irinotecan Metabolite	Oncology	Preclinical	ImaRx Therapeutic Inc
Abraxane	Paclitaxel protein-bound particles for injectable suspension	Paclitaxel-albumin Nanoparticle	Non-small cell lung cancer, breast cancer, others	In the market	Abraxis Bioscience
Cyclosertcamptothecin	linear cyclodextrin-containing polymers	Cyclodextrin nanoparticle	Metastatic solid tumors	Phase 1	Insert Therapeutics
TNT AntiEpCAM	Anti-epithelial cell adhesion molecule mAb	Polymer-coated iron oxide	Solid tumors	Preclinical	Triton BioSystems
Verigene platform	Gold nanoparticle biomolecule probes	DNA-functionalized gold nanoparticles	Diagnostics	In the market	Nanosphere
INGN-401	FUS1 tumor suppressor gene liposome for injection	Liposome	Metastatic lung cancer	Phase 1	Introgen
Combidex	Ferumoxtran-10	Superparamagnetic iron oxide nanoparticle	Tumor imaging	NDA filed	Advanced Magnetics

Nanoxel	Cremophor free soluble formulation	Nano-based drug delivery system for anticancer drug pacilitaxel	Breast, ovarian cancer	In the market	Dabur Pharma
MRX-815	phospholipid -coated nano-bubble	Nanobubbles	Vascular thrombosis	IND filed	ImaRx Therapeutic Inc
Definity	Perflutren lipid microspheres	Lipid-encapsulated octofluoropropane nanospheres	Echocardiogram	Approved	ImaRx/BMS
Doxil	Doxorubicin-HCL liposome injection	Pegylated liposome	Metastatic ovarian cancer	Approved	OrthoBiotech
AmBisome	Amphotericin B liposome for injection	Liposome	Fungal infections	Approved	Gilead Sciences
Rapamune	Sirolimus	Nanocrystalline drug	Immunosuppressant for kidney transplantation	Approved	Elan/Wyeth
Emend	Aprepitant	Nanocrystalline drug	Nausea	Approved	Elan/Merck
NanoDOX®	Topical doxycycline monohydrate hydrogel	1% Doxycycline Monohydrate Hydrogel	chronic inflammatory diseases like diabetic foot ulcers	Phase 2	Nanotherapeutics

Abbreviations: IND = investigational new drug; mAb = monoclonal antibody; NDA = new drug application.

for drug discovery to predict *in vivo* drug response and to minimize animal-based experiments. The advanced methods and screening technologies in drug discovery screening are given below.

ADVANCED METHODS IN CELL-BASED DRUG DISCOVERY SCREENING

a) High Content Analysis
- GPCR Assays
- Viability Assay
- Membrane Integrity Assay
- Apoptosis and necrosis assay
- Caspase activity assay

b) High Throughput Screening
c) Functional Genomics Techniques for Target Discovery
d) Applied Mass Spectrometry
e) ADME

Assay and Screening Technologies

a) Label-Free Screening
b) Fragment-Based Screening
c) Stem Cell Screening
d) Nano Screening
e) Scaffold-Based Approach
f) Application of Epigenetics

Despite of the wide range applications, nanomedicines were mostly used for detecting the targeted biomolecule, drug delivery systems, emulsions, carriers for delivering vaccines and as nanofabricated biomaterials with unusual properties of strength, hardness, reduced friction, and improved biocompatibility. In the last era, less emphasis was given on the use of nanomedicine in drug screening.

In recent years, a generation of numerous hits (drug analogs) in the process of drug discovery screening have triggered a technical revolution within assay development and high throughput screening (HTS) (Wlodkowic et al. 2009). Due to the higher cost of reagents in HTS assays miniaturization of biochemical assay formats were established. The demands for sensitive readouts in HTS have driven the introduction of nanomedicine technologies and fluorescence-based readouts in the process of drug discovery screening. During the past decade, fluorescence assay technology has grown very fast in drug screening based on the receptor-binding approach. Nanomedicine technologies especially quantum dots were used for screening candidate drugs as an alternative to organic fluorophores.

A successful drug candidate binds to the targeted biomolecules without it's off target effect. Quantum dots were used to test such drug candidates as they can be easily tagged with the cellular biomolecules and can be efficiently monitored with different colors of fluorescent emissions. A good hit (drug) might displace blue, aqua and green nanocrystals from the targeted cells but donot displace red, yellow and orange nano-crystals from the non-targeted cells to avoid the side effects (Klarreich 2001). Mattheakis et al. (2004) reported the application of quantum dots for drug screening by tagging different types of cells with different-colored quantum dots. Thus, different-colored quantum dots can be used to tag desired biomolecules and multiple targets can be screened simultaneously with the help of Multi-target High-throughput Screening (MTHTS) (Palfreyman et al. 2002). Quantum dot multiplexing assay can enhance the screening throughput in terms of rapidness, efficiency, stability and accuracy.

Nanomedicines can also be better exploited in high content drug discovery screening assays. Different fluorescent materials and probes are available to support many HCS applications. To date nanomedicine technologies are less commonly used in the process of drug discovery screening. Nano-materials, e.g., QDs are nono-particles with a semiconductor core that has large absorption and narrow emission spectra. Compared to organic fluorophores (immunoreagents, organic dyes or fluorescent proteins), QDs are resistant to photo-bleaching and can be better exploited for HCS applications especially for long-term, highly sensitive live cell imaging (Jan et al. 2008). However, the use of QDs for high content screening has been limited as a result of ineffective delivery into cells.

A few research articles report the application of nanomedicine technologies for drug screening. A quantum dot-labeled ligand-receptor binding assay for G protein-coupled receptors (GPCR) contained in minimally purified membrane nanopatches has been reported by Swift et al. (2009). This assay relies on the ligand and receptor interactions and can be used in the process of drug discovery screening. Rohde et al. (2008) reported a high-throughput *in vivo* genetic and drug screening using femtosecond laser nano-surgery and microfluidics. Yun et al. (2003) described a micro/nano-fluidic chip-based micro-well array for high-throughput drug screening. Similarly, Vo-Dinh (1998) described a nanoprobe for surface-enhanced Raman spectroscopy in medical diagnostic and drug screening. Nanoprobe has a spherical shape with a diameter less than one micrometer and preferably in the 10–500 nanometer range. Nanoporous, biocompatible, alumina membranes as a platform for incorporation into a cell based device targeted for in situ recording of cellular electrical activity variations due to the changes associated with the surrounding microenvironments more specifically due to the effect of therapeutic drugs was reported by Prasad and Quijano (2007). This will serve as a nano-structured biomedical drug testing device.

Tunneling Nano Tube (TNT) based drug screening system has been designed by a BTO company, Bergen, Norway (http://www.uib.no/tto/technologies/tnt.pdf). The system can be used in the development of lead compounds for cancer treatment, multi-drug resistance (e.g., p-glycoprotein), viral spreading, autoimmunity, etc. Micro- and nanofluidic systems for high-throughput biological screening have been reported Hong et al. (2009). This miniaturized system can elucidate the alterations in the intracellular bio-molecules due to the effect of drug analogs in a high-throughput format. The automated Nano-Patch- Clamp can enable high quality and high content ion channel and compound evaluation on a one-cell-at-a-time basis.

We describe simultaneous quantitative monitoring of drug-induced caspase cascade pathways in carcinoma cells using high-content cellular imaging cytometry (Naoghare et al. 2010). The developed platform is compatible for drug screening using commercially available fluorophores, substrates and quantum dots.

In the process of drug discovery screening false-positive hits often interfere with new molecular targets (Wlodkowic et al. 2009). Selecting the appropriate method is necessary to for drug development work. Cellular assays like caspase activation/inhibition and apoptotic assays have been used for drug discovery screening (Lai et al. 2003). The activation of caspases involves complex biochemical pathways. This in turn limits the development of new modalities against caspases as therapeutic targets. In biochemical assays large numbers of cells need to be lysed in order to measure the caspase activity. This facilitates the entire caspase activity measurement but not the intercellular variation in the caspase activity within a population. Measurement of intercellular variations in the caspase activities may differentiate a damaged cell from the normal cell. Thus, it is important to measure the caspase activities in individual cells. It is important to note that apart from the crucial role in apoptotic mechanism, caspases are also involved in inflammation. Activation of caspases does not always leads to apoptosis. Therefore, it is important to simultaneously monitor the initiator (caspase-2 and 8) and downstream (caspases-3) caspases to elucidate the drug effect.

A high-content cellular imaging cytometry for the simultaneous quantitative monitoring of caspase activation has been developed. With this various cellular caspases involved in apoptotic pathways can be targeted for their possible role as therapeutic targets.

Human myeloid leukemia (HL-60) cells were treated with SH-03 {(7S,7aR,13aS)-9,10-dimethoxy-3,3-dimethyl-7,7a,13,13a-tetrahydro-3H-chromeno [3,4-b]pyrano[2,3-h]chromen-7-ol} (a newly synthesized candidate), camptothecin or naringenin. SH-03 is a newly synthesized drug analog, whereas camptothecin and naringenin are known to induce apoptosis in myeloid cells (Naoghare et al. 2010). Rapid and quantitative analysis of drug-induced caspase cascades was achieved by a newly developed approach of uniform threshold intensity distribution. Drug-induced caspase cascades were monitored through multi-spectral and multi-color imaging cytometry. SH-03 triggered the apoptotic signals in HL-60 cells through caspase-9-mediated (intrinsic) apoptotic pathway by inhibiting the PI3K/AKT signal transduction pathway. Qdot based apoptotic assay was used for validation of the obtained results. We hope that the high-content cellular imaging cytometry can efficiently be used in the process of drug discovery screening.

METHODOLOGY

Cell Culture and Drug Treatment

HL-60 cells were maintained in RPMI-1640 medium with appropriate culture and environmental conditions. Cells (4.5 x 10^5 cells/ml) were treated with SH-03, camptothecin (CAM) or naringenin (NAR). Before the drug treatment, cells were grown overnight so that they were in log-phase growth. Three independent experiments were performed for each particular drug concentration at a given time interval to achieve better statistical confidence.

APOPTOSIS ASSAY THROUGH QUANTITATIVE IMAGING CYTOMETRY AND FLOW CYTOMETRY

Imaging Cytometry

After drug treatment, cells were washed and resuspended in 100 µl of calcium-enriched binding buffer (BD Biosciences, CA, USA) to allow annexin V binding to PS (phosphatidylserine) on the cellular membrane. To the cell suspension 5 µl of annexin V-Biotin (BD Biosciences) was added followed by incubation at room temperature for 15 min in the dark. Unbound annexin V-Biotin was removed by the washing step. 100 µl of binding buffer was added to the cell pellet, cells were resuspended and incubated with 10 nM Qdot-streptavidin conjugate (Cat No. Q10141MP, Invitrogen, CA, USA) and 10 µl of propidium iodide (BD Biosciences) for 15 min in a dark at room temperature. Finally, the cells were washed and resuspended in 10 µl of binding buffer and analyzed on quantitative cellular imaging analysis. To avoid possible aggregation quantum dots were dissolved in binding buffer and vortexed before the staining procedure.

Flow Cytometry

To validate the results obtained by quantitative imaging cytometry SH-03 treated cells were analyzed on flow cytometry. To monitor necrotic cell death propidium iodide (PI) was used, whereas the apoptotic cells were monitored by fluorescein isothiocyanate (FITC)-labeled annexin-V (Apoptosis Detection Kit; BD Biosciences, Franklin Lakes, NJ, USA). Cell pellets containing approximately 1×10^5 cells were resuspended in 100 µl of calcium-enriched binding buffer. To the resuspended cell culture, 5 µl FITC-labeled annexin V and 5 µl PI was added followed by incubation for 10 min at 4.0°C in the dark. After incubation, 400 µl of binding buffer was added and the cells were analyzed on FACSCalibur apparatus (BD Biosciences).

Caspase-3, caspase-8 and caspase-9 activation assays

Activation of initiator and downstream caspases were monitored using quantitative cellular imaging cytometry. Different fluorescent substrates i.e., caspase-3 [(z-DEVD)2-Magic Red], caspase-8 [(Z-IETD)2 -Rh110] and caspase-9 [(Ac-LEHD)2-Rh 110] were used to monitor the caspase activities. Substrates were purchased from Immunochemistry Technologies, LLC, Bloomington, MN and AnaSpec, Inc., San Jose, CA. As per the manufacturers' recommendations stock solutions of caspase-3, caspase-8 and caspase-9 substrates were prepared in DMSO. Cells were treated with the respective drug. After the drug treatment cells were washed and resuspended in complete RPMI-1640 medium with 40 µM final concentration of caspase-9 or caspase-8 substrate and incubated at 37°C in the dark for 2 hr. After the incubation, cells were washed and resuspended in RPMI-1640 medium (without serum) with 20 µM final concentration of caspase-3 substrate and incubated at 37°C in the dark for 1 hr. To avoid the possible interference of serum proteins with the fluorescent caspase substrates, serum-free medium was used during imaging analysis. Cells were excited at 488 nm with an Ar-ion laser and emissions were recorded over the entire spectral range from 463 nm to 688 nm. The emission maxima for caspase-3, caspase-8 and caspase-9 substrates were recorded at 617, 523 and 523 nm, respectively.

Caspase-8 and Caspase-2 inhibition Assays

Caspase-8 and caspase-2 inhibition assays were carried out to elucidate mechanistic insights of the drug-induced apoptosis. Z-IETD-FMK (20 µM; BD Pharmingen, CA, USA) was used as an inhibitor of caspase-8, whereas Z-VDVAD-FMK (20 µM; Calbiochem, CA, USA) was used as a caspase-2 inhibitor. Cells were pre-incubated with a respective inhibitor for 15 min followed by the drug treatment (SH-03, CAM or NAR).

ELECTROPHORESIS AND WESTERN BLOTTING

Cells (1×10^6) were treated with 10 µM SH-03 for 12 hr, washed with PBS and lysed. Lysis-M solution (Roche, Manheim, Germany) was used to prepare the whole cell extracts. Cell lysates containing 60 µg protein were loaded on 12% sodium dodecyl sulfate-polyacrylamide gel and electrophorized. Gel was transferred to a polyvinylidene difluoride membrane (BioRad, Hercules, CA, USA), and hybridized with an appropriate primary antibody (anti- cleaved caspase-8; anti-cytochrome C; anti-alpha tubulin; Cell Signaling Technology, Waltham, MA, USA). Western blots were blocked in PBS containing Tween 20 (0.05%) and bovine serum albumin (5%) at 4°C over night. Primary antibodies were detected using a 1:10,000 dilution of secondary peroxidase-conjugated antibody (Dako Ltd., High Wycombe, Bucks, UK). The membrane was incubated for 2hr at room temperature followed by washing with PBS-Tween. Signals were developed with an enhanced chemiluminescence detection kit (Roche, Basel, Switzerland).

PI3K/AKT ACTIVATION AND PHOSPHORYLATION

The effect of SH-03-induced intrinsic apoptotic pathway on the PI3K/AKT activation/phosphorylation was assayed. HL-60 cells were treated with 10 µM of either SH-03 or with the specific PI3K inhibitor LY294002 (Cell Signaling Technology) for 12 hr. The cells were washed and fixed with 4% 1-ethyl-3-(3-dimethylaminopropyl)carbodiimide hydrochloride for 15 min, and permeabilized with 0.2% saponin. Primary antibodies, i.e., rabbit anti-PI3K and mouse anti-AKT (Cell Signaling Technology) or rabbit anti-p-PI3K (Santa Cruz Biotechnology, Santa Cruz, CA, USA) and mouse anti-p-AKT (Cell Signaling Technology) were used for labeling with 1:100 dilution ratio. Cells were washed with PBS and labeled with secondary antibody (goat anti-rabbit IgG-peridinin-chlorophyll-protein complex (PerCP) or goat anti-mouse IgG-FITC; Santa Cruz Biotechnology) at a 1:30 dilution for 1hr at room temperature. Cells were finally subjected for analysis using quantitative cellular imaging cytometry. Cells were excited at 488 nm and the emission maxima for AKT, pAKT, PI3K, and pPI3K were collected at 525, 525, 675, and 675 nm, respectively.

QUANTITATIVE HIGH-CONTENT CELLULAR IMAGING CYTOMETRY

Data acquisition and quantitative analysis using high-content cellular imaging cytometry has been described in detail in our earlier publication (Naoghare et al. 2008). Briefly, cells were treated with drugs and mounted on the sample stage. Cells were excited with 488-nm beam from an Ar-ion laser (Melles Griot Laser Group, 35-LAP-431-220). The desired Bragg angle (θ) was obtained using a bi-refringent crystal built into the acousto-optic tunable filter (AOTF). CCD camera was used to detect the fluorescence beam transmitted from the AOTF. Hyper-spectral images were taken within the spectral region from 463 nm to 688 nm with a 3.75 nm interval and a scanning rate of 1 wavelength/second. Image analysis was carried out using commercially available software (MetaMorph, Version 7.1.3.0, Molecular Devices). Schematic diagram of the quantitative high-content cellular imaging cytometry is described in Fig. 7.3.

RESULTS

SH-03 Triggers Apoptosis in HL-60 Cells

Figure 7.3B shows our generalized approach of the simultaneous monitoring of the effect of the drug on cellular biomolecules. In it, inhibition of the PI3K/AKT signaling pathway the induction of caspase-9/caspase-3 by SH-03 can be observed. Chemical structure of SH-03 has been shown in Fig. 7.1B. SH-03 induced apoptosis in HL-60 cells in a dose-dependent manner. Cellular images were obtained as a function of caspase substrate reaction at different cellular emissions via coincident operation of AOTF scanning and CCD imaging. SH-03 concentration was fixed to 10 μM through DNA fragmentation assay. SH-03 was found to trigger apoptosis in HL-60 cells after 6 hr of treatment. Cellular images of annexin V-Qdot- and propidium iodide (PI)-stained cells were taken at 523 nm and 617 nm, respectively. Bright fluorescent ring around the cell membrane at 523 nm can be seen due to the staining to PS moieties by Annexin V-Qdot, whereas necrotic cells were observed at 617 nm at emission maxima of PI. The super-imposed image was taken at two different wavelengths, i.e., 523 nm and 617 nm. These results were verified using flow cytometric analysis. Scatter graphs (FSC-H/SSC-H) and the dot plots of HL-60 cells showed the induction of apoptosis (32.5%) after 6 hr of SH-03 (10 μM) treatment.

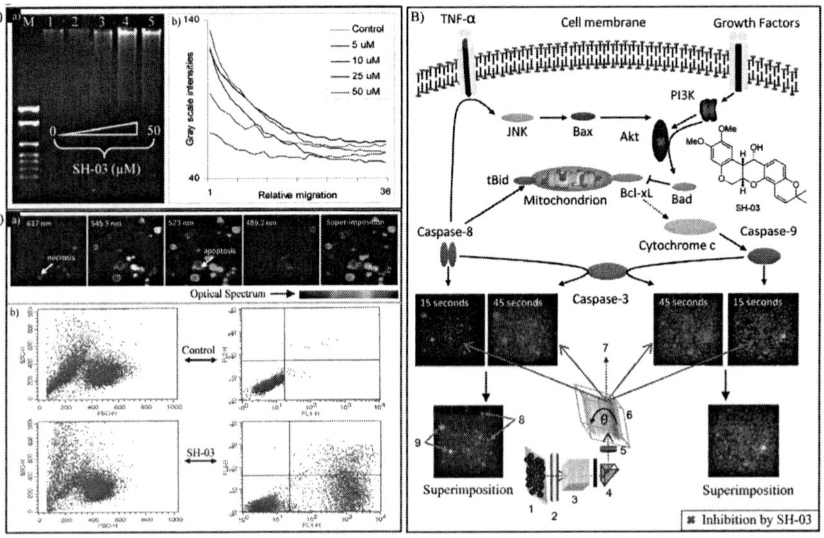

Figure 7.3: (A) Naoghare et al. 2010. DNA fragmentation in HL-60 cells after SH-03 treatment. Cells were treated with different concentrations of SH-03 and incubated for 6 hr at 37°C. (a) DNA was isolated from the cells and electrophorized on 0.8% agarose gel. M signifies the marker lane (1 kb DNA ladder), and lanes 1–5 denotes results obtained using 0, 5, 10, 25, and 50 μM SH-03, respectively. (b) Line graph depicting the average gray scale intensities of each DNA sample. (B) Structure of SH-03 [(7S,7a*R*,13a*S*)-9,10-dimethoxy-3,3-dimethyl-7,7a,13,13a-tetrahydro-3*H*-chromeno [3,4-b] pyrano[2,3-h]chromen-7-ol] and the activation of caspase-9 and caspase-3 after SH-03 treatment. Hyperspectral fluorescent images (SH-03; 10 μM) obtained using cellular imaging cytometry. In it, 1. Sample, 2. Objective lens (60 X), 3. Beam splitter, 4. Prism, 5. Focusing lens, 6. AOTF, 7. Un-diffracted beam, 8. Caspase-3 activation, 9. Caspase-3 and caspase-9 activation. θ. Bragg angle (the angle between diffracted and un-diffracted beam). (C) Representative hyperspectral fluorescent images of SH-03 (10 μM, 6 hr) induced apoptosis in HL-60 cells. (a) Images were obtained over a spectral region from 463–688 nm with 60s of total time interval. Images taken at 523 nm and 617 nm were super-imposed. (b) Flow cytometric analysis illustrating the scatter and dot plots of HL-60 cells after 6 hr of SH-03 treatment. The lower right quadrant represent annexin V-FITC positive cells, whereas upper left quadrant represents propidium iodide (PI)-positive dead cells (necrosis).

Color image of this figure appears in the color plate section at the end of the book.

SH-03 Triggers Death Signals via Intrinsic Apoptotic Cascade

SH-03 was observed to induce caspase-9 and caspase-3 activation in HL-60 cells. Caspase-9 activation was observed in 5.39, 72, 79.28 and 79.90% of control cells and cells treated with 5, 10 and 25 μM SH-03, respectively. Caspase-3 activation in those cells was 6.86, 76, 93.49, and 85.78%, respectively. Caspase-2 inhibitor, Z-VDVAD-FMK was used to demonstrate the involvement of caspases in SH-03-induced apoptosis. Our results showed that the pretreatment with caspase-2 inhibitor (20 μM) had reduced SH-03-induced caspase-9 and caspase-3 activation. From this it is inferred that SH-03 triggered caspase-9 mediated apoptotic pathway in HL-60 cells. Caspase-8 and caspase-3 activation assays were carried out to check the involvement of SH-03-induced apoptosis through an extrinsic apoptotic pathway. Percent caspase-8 activation was 1.44, 6.83, 8.53 and 10.10% in control cells and in cells treated with 5, 10 and 25 µM SH-03, respectively. Whereas, the percent activation of caspase-3 was 5.25, 35.93, 50.12 and 65.61% in these cells, respectively. The obtained results showed that SH-03 does not trigger caspase-8 activation. Further validation was carried out using a caspase-8 inhibitor (Z-IETD-FMK). Pretreatment with caspase inhibitor (20 µM) did not significantly change SH-03-induced apoptosis. Further validation was carried out using Western blot analysis. Cytosolic cytochrome C was found to be elevated in SH-03 treated cells, whereas the caspase-8 activity remained the same in control as well as in drug treated cells. The level of alpha tubulin (positive control) was observed to be the same in treated and untreated cells.

Effect of SH-03 and LY294002 on PI3K/AKT Activation and Phosphorylation

PI3K/AKT activation and phosphorylation assays were carried out to get the mechanistic insights of the SH-03-induced intrinsic apoptotic pathway in HL-60 cells. A known and specific inhibitor of PI3K (LY294002) was used to monitor the AKT, pAKT, PI3K, and pPI3K activities in HL-60 cells. AKT, pAKT, PI3K, and pPI3K activities in control was found to be 80.5%, 69.9%, 84.7%, and 74.7%, respectively, compared to 13.6%, 5.3%, 12.1%, and 7.3%, respectively, in LY294002 treated cells. On the other hand, AKT, pAKT, PI3K, and pPI3K activities in SH-03 treated cells were 6.8%, 8.4%, 3.9%, and 7.9%, respectively. These findings suggest that SH-03 blocked the PI3K/AKT signaling pathway by inhibiting the activation and phosphorylation of PI3K and AKT. Furthermore, treatment with SH-03 efficiently inhibited AKT, pAKT, PI3K, and pPI3K activities in HL-60 cells compared to LY294002-treated cells.

CAM AND NAR INDUCE ACTIVATION OF CASPASE CASCADES

To validate our newly developed method, agents (CAM and NAR) known to initiate caspase activation were used. Treatment with CAM induced caspase-9 activation in HL-60 cells. Pretreatment with caspase-2 inhibitor resulted in the activation of caspase-9 in 8.40, 12.60, 19.37 and 33.35% of control cells and cells treated with 50, 100 and 287 nM CAM, respectively. Caspase-3 was found to be activated in 12, 47.27, 52.30 and 64.90% of these cells, respectively. Treatment with 0.1, 0.25 and 0.5 mM NAR activated Caspase-9 in 8.27, 14.60 and 20.87% cells compared to the control (5.9%).

Pretreatment with caspase-2 inhibitor resulted in caspase-3 activation in 5.90, 6.73, 17.87 and 21.10% in control and 0.1, 0.25 and 0.5 mM NAR treated cell. These results suggest that either CAM or NAR triggers death signal via an intrinsic apoptotic pathway. The possibility of CAM and NAR induced extrinsic apoptotic pathway was also examined. CAM was found to induce caspase-8 and caspase-3 activation in a dose-dependent manner, whereas the treatment NAR did not significantly induce caspase-8. These results suggest that CAM triggers the apoptotic caspase cascade in HL-60 cell through caspase-9 and caspase-8 activation, whereas NAR was found to trigger the caspase cascade through caspase-9 activation.

Discussion

Recently, new therapeutic approaches were developed to elucidate complex apoptotic signaling pathways (Lai et al. 2003). Initially, a few classes of drug target molecules were focused using the conventional therapeutic approaches. Better understanding of new drug targets (e.g., initiator and effector caspases) are needed to elucidate complex apoptotic mechanism and to design appropriate therapeutic approaches. Researchers are now interested in investigating the role of caspases as a therapeutic target considering their important role in various diseases (e.g., Alzheimer's, Huntington and other neurodegenerative diseases) (McBride et al. 1999). Broad-spectrum caspase inhibitors have been clinically evaluated to determine their use in delaying or preventing the progression of various diseases (Zheng et al. 2006). Fewer reports are available on the use of caspase inhibitors in leukemia therapy (Blagosklonny 2000). In this regard, quantitative cellular imaging cytometry can facilitate the elucidation of caspase-mediated cell death dynamics through caspase activation and inhibition assays. In the present work we have simultaneously monitored the activity of initiator and downstream caspases in HL-60 cells. The developed method allows rapid and automated analysis of the drug-induced cellular response. The results obtained by flow cytometric analysis and cellular imaging cytometry were in line and thus validated our approach. The developed approach strengthens the application of image-based cellular assay in the process of drug discovery screening.

KILLER APPS IN NANOMEDICINE TECHNOLOGIES AND THE FUTURE APPROACH

For centuries, researchers searched for miracle cures to stop the sufferings caused by diseases and injuries. With the advent of nanomedicine technologies new options are now available in selecting better therapeutic targets. Nanomedicine technologies involve the design and manufacture of devices at nanometer level (Morrow et al. 2007). Nanomedicine is divided into three areas, i.e., nano-diagnostics, targeted drug delivery and cell and tissue repair. Nano-diagnostics involves *in-vitro* and *in-vivo* models that are capable of quantifying and visualizing the expressions of targeted biomolecules at cellular level. Targeted drug delivery involves the specific delivery of drug molecules to the targeted cells. Cell and tissue repair deals with the synthetic growth of tissue by targeted delivery. Many government agencies like U.S. National Cancer Institute have invested hundreds of millions of dollars in nanomedicine. U.S. National Nanotechnology Initiative provided more than US$200 million in the 2008 budget for the National Institutes of Health (NNCO 2008). European countries like Germany and the UK are also investing money in nanomedicine research. Recently, new imaging tools have been developed for drug discovery screening applications. But better state-of-the-art equipment is needed for the efficient exploitation of nanomedicines. Among the nanomedicine technologies, quantum dots have been efficiently exploited for real applications in imaging. Nanomedicine technologies have been used for drug delivery and regenerative medicine but they have not been exploited in the field of drug discovery screening. Thus there is a need to develop new modalities for better exploitation of nanomedicine technologies in the process of drug discovery screening.

TOXICOLOGICAL CONCERNS

Researchers around the globe are trying to exploit nanomedicine for the betterment of human health. However, certain toxicological and ethical concerns also need to be addressed along with the benefits of nanomedicine. Lack of knowledge about the nanoparticle toxicity in the *in vitro* and *in vivo* models can create unseen problems (Lee et al. 2009). Despite these concerns, nearly 130 nanotech-based drugs and delivery systems and 125 devices/diagnostic tests have entered pre-clinical, clinical, or commercial development since 2005 (NanoBiotech News 2006). Recently, NIH is working on the safety issues involved in using nanoparticles. This includes the cytotoxicity assessments, length of time period of nanoparticle left in the body, dermal exposure, undesired effects etc.

RESEARCH GAPS

Nanomedicine technologies can manipulate the biological systems at the molecular level. However the success rate is likely to be slower as there are issues related to the ethical and intellectual property rights, side-effects and toxicological concerns of nanomedicine technologies. These issues are hindering the commercial exploitation of nanomedicine technology. In case of drug discovery screening, nanomedicine technologies have not been exploited efficiently. There is an urgent need to develop new modalities for better exploitation of nanomedicine technologies in the process of drug discovery screening.

Secondly, the nanomedicine technologies have been mostly studied for basic research rather than its commercial application. Therefore, the industrial participation for promoting nanomedicine technologies was not significant compared to the participation of the basic research groups. In the coming years, nanomedicine technologies will continue to expand in different areas of life sciences. Nanomedicine technologies will address many important problems with the help of nanoscale-structured materials. Nanomedicine technologies will find their way in various areas including cellular assays, chemotherapy, bio-sensors, bio-chips, needle less injectors etc. Although there are many things which need to be understood, with the combined efforts of physicists, chemists and biologists, nanomedicine technologies will become a standard tool for biological applications.

REFERENCES

Blagosklonny, M.V. 2000. The dilemma of apoptosis in myelodysplasia and leukemia: a new promise of therapeutic intervention? Leukemia 14(12): 2017–2018.

Ebbesen, M. and T.G. Jensen. 2006. Nanomedicine: Techniques, Potentials, and Ethical Implications. J Biomed Biotechnol (5): 51516.

Freitas, R.A. Jr. 2005. Current Status of Nanomedicine and Medical Nanorobotics. J Comput. Theor Nanosci 2: 1–25.

He, J. and X. Qi, Y. Miao, H.L. Wu, N. He and J.J. Zhu. 2010. Application of smart nanostructures in medicine. Nanomedicine (Lond) 5(7): 1129–1138.

Hong, J. and J.B. Edel and A.J. deMello. 2009. Micro- and nanofluidic systems for high-throughput biological screening. Drug Discov Today 14(3-4): 134–146.

Jan, E. and S.J. Byrne, M. Cuddihy, A.M. Davies, Y. Volkov, Y.K. Gun'ko and N.A. Kotov. 2008. High-content screening as a universal tool for fingerprinting of cytotoxicity of nanoparticles. ACS Nano 2(5): 928–938.

Klarreich, E. 2001. Biologists join the dots. Nature 413: 450–452.

Lai, J-M. and C-H. Hsieh and Z-F. Chang. 2003. Caspase activation during phorbol ester-induced apoptosis requires ROCK-dependent myosin-mediated contraction. J Cell Sci 116: 3491–3501.

Lee, J. and G.D. Lilly, R.C. Doty, P. Podsiadlo and N.A. Kotov. 2009. *In vitro* toxicity testing of nanoparticles in 3D cell culture. Small 5(10): 1213–1221.

Lunn, C.A. 2010. Label-free screening assays: a strategy for finding better drug candidates. Future Med Chem 2(11): 1703–16.

Mattheakis, L.C. and J.M. Dias, Y.-J. Choi, J. Gong, M.P. Bruchez, J. Liu and E. Wang. 2004. Optical coding of mammalian cells using semiconductor quantum dots. Anal Biochem 327(2): 200–208.

McBride, C.B. and L.T. McPhail and J.D. Steeves. 1999. Emerging therapeutic targets in caspase-dependent disease. Expert Opin Ther Targets 3(3): 391–411.

Morrow, K.J. Jr. and R. Bawa and C. Wei. 2007. Recent advances in basic and clinical nanomedicine. Med Clin North Am 91(5): 805–843.

Naoghare, P.K. and M.J. Kim and J.M. Song. 2008. Uniform threshold intensity distribution-based quantitative multivariate imaging cytometry. Anal Chem 80: 5407–5417.

Naoghare, P.K. and H.A. Ki, S.M. Paek, Y.K. Tak, Y.G. Suh, S.G. Kim, K.H. Lee and J.M. Song. 2010. Simultaneous quantitative monitoring of drug-induced caspase cascade pathways in carcinoma cells. Integr Biol (Camb) 2(1): 46–57.

Palfreyman, M.G. and D.J. Hook, L.J. Klimczak, J.A. Brockman, D.M. Evans and C.A. Altar. 2002. Novel directions in antipsychotic target identification using gene arrays. Curr Drug Targets CNS Neurol Disord 1(2): 227–238.

Patil, M. and D.S. Mehta, and S. Guvva. 2008. Future impact of nanotechnology on medicine and dentistry. J Indian Soc Periodontol 12(2): 34–40.

Prasad, S. and J. Quijano. 2007. Nanoporous noninvasive cellular electrical activity-based analysis devices. Clin Lab Med 27(1): 75–91.

Rohde, C. and C. Gilleland, C. Samara, F. Zeng and M.F. Yanik. 2008. High-throughput *in vivo* genetic and drug screening using femtosecond laser nano-surgery, and microfluidics. Conf Proc IEEE Eng Med Biol Soc 2008: 2642.

Saha, M. 2009. Nanomedicine: Promising Tiny Machine for the Healthcare in Future. OMJ 24: 242–247.

Special report details rapid growth of nanomedicines, diagnostics pipeline. January 4, 2006. NanoBiotech News. http://www.caltechcancer.org/filestore2/download/62/Nanobiotech%20News%201-4-06.pdf

Swift, J.L. and M.C. Burger and D.T. Cramb. 2009. A quantum dot-labeled ligand-receptor binding assay for G protein-coupled receptors contained in minimally purified membrane nanopatches. Methods Mol Biol 552: 329–341.

The National Nanotechnology Initiative. 2008. National Nanotechnology Coordination Office (NNC0). U.S. Government. http://www.nano.gov/NNI_08Budget.pdf

Tunneling Nano Tube (TNT) based drug screening system. http://www.uib.no/tto/technologies/tnt.pdf

Vo-Dinh, T. 1998. Nanoprobe for surface-enhanced Raman spectroscopy in medical diagnostic and drug screening. United States Patent No. 6219137.

Wee, K.W. and G.Y. Kang, J. Park, J.Y. Kang, D.S. Yoon, J.H. Park and T.S. Kim. 2005. Novel electrical detection of label-free disease marker proteins using piezoresistive self-sensing micro-cantilevers. Biosens Bioelectron 20(10): 1932–1938.

Wlodkowic, D. and J. Skommer, D. McGuinness, S. Faley, W. Kolch, Z. Darzynkiewicz and J.M. Cooper. 2009. Chip-based dynamic real-time quantification of drug-induced cytotoxicity in human tumor cells. Anal Chem 81: 6952–6959.

Yun, K-S. and E. Yoon. 2003. A micro/nano-fluidic chip-based micro-well array for high-throughput cell analysis and drug screening. 7th lnternational Conference on Miniaturized Chemical and Blochemlcal Analysts Systems. October 5–9, 2003, Squaw Valley, Callfornla USA.

Zheng, C.J. and L.Y. Han, C.W. Yap, Z.L. Ji, Z.W. Cao and Y.Z. Chen. 2006. Therapeutic targets: progress of their exploration and investigation of their characteristics. Pharmacol Rev 58: 259–279.

Zheng, G. and F. Patolsky, Y. Cui, W.U. Wang and C.M. Lieber. 2005. Multiplexed electrical detection of cancer markers with nanowire sensor arrays. Nat Biotechnol 23(10): 1294–1301.

8

Emerging Nanotechnology for Efficient Capture of Circulating Tumor Cells

Xueli Liu,[1] *Li Chen*[1] and *Shutao Wang*[1,a,*]

INTRODUCTION

Metastasis is a common biological process of solid epithelial tumors. This process was vividly described as the 'seed and soil' theory by Stephen Paget in 1889 (Paget 1889). During tumor metastasis, tumor cells shed from the primary tumor and invade the surrounding tissue, enter either the blood or the lymphatic circulation, extravasate and finally grow at a distant tissue (Steeg 2006). Metastasis is the main cause of death for cancer patients. It is necessary to develop an effective method to monitor and treat tumor metastasis, as traditional treatments such as surgery and radiographic imaging aim at the primary tumor, and chemotherapy and hormonal therapy usually act as palliative care.

Tumor cells that survive and circulate in the peripheral blood during the metastasis shedding from either the primary tumor or its metastases are called circulating tumor cells (CTCs). The detection of CTCs is very important for early diagnosis of metastases, cancer prognosis, personalization treatment and the biological study of metastases. Many methods have been developed, in which immunologic techniques

[1]Beijing National Laboratory for Molecular Sciences (BNLMS), Institute of Chemistry, Chinese Academy of Sciences, Beijing, 100190 (P. R. China).
[a]E-mail: stwang@iccas.ac.cn.
*Corresponding author

and PCR-based assays (Pantel et al. 1999, Schoenfeld et al. 1997) are the most commonly used ones. Immunologic techniques are based on immunostaining with monoclonal antibodies against epithelium-specific antigens to identify CTCs. Cytokeratins are the commonly used protein makers. PCR-based assays analyze the expression of genes specific to CTCs by reverse transcription targeting RNA markers. Immunocytochemistry facilitates subsequent morphological identification while the latter has higher sensitivity (Datta et al. 1994, Riethdorf et al. 2007, Smith et al. 2000). However, as CTCs are very rare in the peripheral blood (about a few CTCs compared to approximately 10 million leukocytes and five billion erythrocytes in the blood), an capture and enrichment process is needed ahead rather than detection directly.

Huge efforts have been developed to explore new approaches for CTCs capture since CTCs were detected and firstly reported by T.R. Ashworth (Ashworth 1869). Along with the evolution of nanotechnology, the combination of biology, medicine and micro- and nano- fabrication technologies broadens the field of drug synthesis, molecule analysis and disease treatment (Dittrich and Manz 2006, Hong and Quake 2003, Rosi and Mirkin 2005, Yager et al. 2006). Many new methods of CTCs enrichment are designed with emphasis on the microscale interaction between cells and the substrates. In this chapter, we will focus on these different enrichment methods of CTCs especially the new emerging ones and discuss the prospect of CTCs isolation and detection.

EMERGING METHODS OF CTCs CAPTURE

As CTCs are rare in the peripheral blood, contaminant of leukocytes, erythrocytes and platelets will reduce the sensitivity and specificity of CTC detection. Therefore a series of enrichment and isolation methods have been developed. Traditional methods including morphology based separation and immunomagnetic selection are widely used by laboratory and clinical research, several new methods such as microfluidic device based separation and nanoscaled substrate based CTCs capture also have their application value. We will focus here on these methods and their recent advances, potentials and limitations.

Morphology-based Methods

Morphology-based enrichments include filtration and density gradient centrifugation. The former relies on the larger size of tumor cells compared to hemocytes, while the latter utilizes density differences of cells.

ISET (Isolation by Size of Epithelial Tumor Cells) which makes use of the same principle with Nucleopore assay isolates tumor cells individually by using porous filter membrane with 8 μm pores (Ntouroupi et al. 2008, Pinzani et al. 2006, Vona et al. 2000). The separated cells are then fixed and stained for the next analysis. Recently a parylene membrane microfilter device in combination with MEMS (micro-electro-mechanical system) (Zheng et al. 2007) was used to capture CTCs mixed in human blood with ~ 90% recovery, with the biggest advantage of electrolysis cells in situ which shortened the subsequent CTCs identification time (Fig. 8.1). The filtration method has the advantages of a high enrichment rate along with an easy manipulation process, but the disadvantages include contamination and CTCs missing that concern cell heterogeneity as well as non-specificity.

Density-gradient centrifugation separates mononuclear cells including CTCs from blood cells and granulocytes based on their density differences using a density gradient of 1077g/mL. Ficoll-Hypaque and Lymphoprep are the two commonly used density-gradient solutions. This technology is feasible and time saving. However, the loss of tumor cells can occur due to the migration of cells to the plasma layer below and the sediment of cell to the bottom of the centrifuge tube. What's more, the centrifugation process must be developed immediately to prevent the mixing of different layers. In order to overcome these drawbacks, Oncoquick (Müller et al. 2005) was developed by using a porous layer to prevent contamination and cell loss caused by layer mixing and cell sediment.

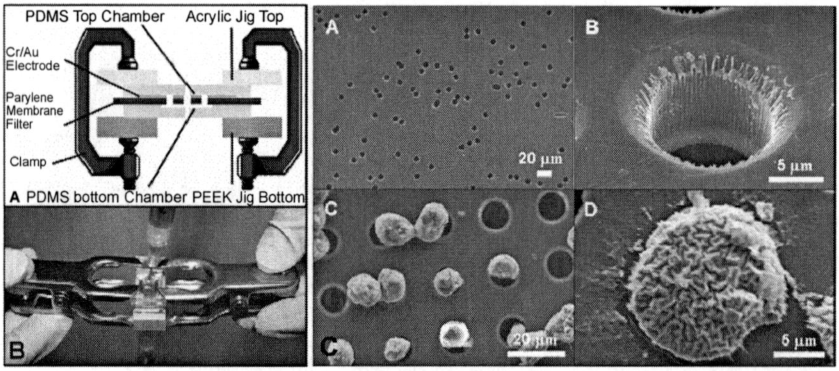

Figure 8.1: Schematic of the membrane filter device. A) Illustration of the assembly device. The parylene membrane sandwiched between PDMS pieces and clamped by the jig. B) A picture of the device. C) SEM pictures of the capture of cells. A, the filter membrane; B, magnified SEM image of the pore in the filter membrane; C, captured cells on the membrane without SEM fixation treatment; D, magnified SEM image of C after SEM fixation procedure.

Immunomagnetic-based Selection

Immunomagnetic selection is a specific antigen-antibody identification and subsequent physical separation process (Bilkenroth et al. 2001, Flatmark et al. 2002, Naume et al. 1997, Naume et al. 1998). In brief, CTCs in whole blood are identified and selected by antibodies tumor-specific molecule coupled to magnetic particles (beads or ferrofluids) targeting epithelial markers (such as CKs or EpCAM), and then separated from other blood cells by a magnetic field to attract and fix the magnetic particles (Racila et al. 1998). Sometimes a negative selection is carried out by using magnetic beads coupled with anti-CD45 or anti-CD61 antibody, or a pan-leukocyte marker. Some practical techniques such as the MAS (Magnetic Activated Cell Sorting System) (Martin et al. 1998) or Dynal Magnetic Beads have been developed, in which the FDA-approved Cell Search System is the most mature one. The Cell Search System is a semi-automated analyzer enriching CTCs by first positive selection using ferrofluids coated with anti-EpCAM antibodies, then negative selection to get rid of lecocytes, staining and finally identification using a four color fluorescence microscope. Immumomagnetic selection is specific and more sensitive than morphology based methods (Balic et al. 2005). However, the lack of a tumor marker for all CTCs may cause CTCs missing as actually not all CTCs will express related antigens or surface proteins at the same time (Goeminne et al. 2000). Moreover, non-epithelial cells can express the epithelial markers thus causing false positive selection.

With the boom of nanomaterial and nanotechnology, the combination of microfluidic system and magnetic materials including magnetic particles has drawn a lot of research attention for the potential application in biological cells handling and separation, magnetic nucleic acid assays, catalysis, and so on (Gijs et al. 2010). For example, a glass microfluidic chip was used to separate red blood cells relying on their native magnetic properties rather than labeling with a high capture rate (93.7%) by using a nickel wire for generating the magnetic gradient inside the channel (Qu et al. 2008). In another work (Adams et al. 2008b), a multitarget magnetic activated cell sorter (MT-MACS) combined with microfluidic technology was used to select two bacterial cell types simultaneously and separately out of the mix which contained a third non-target cell in it (Fig. 8.2). Two magnetic tags with distinct saturation magnetization and size separately coupled with different target-specific affinity reagents were first mixed with the sample, and then separated by magnetophoresis when passed through the microchannel with 90% purity and > 5000-fold enrichment.

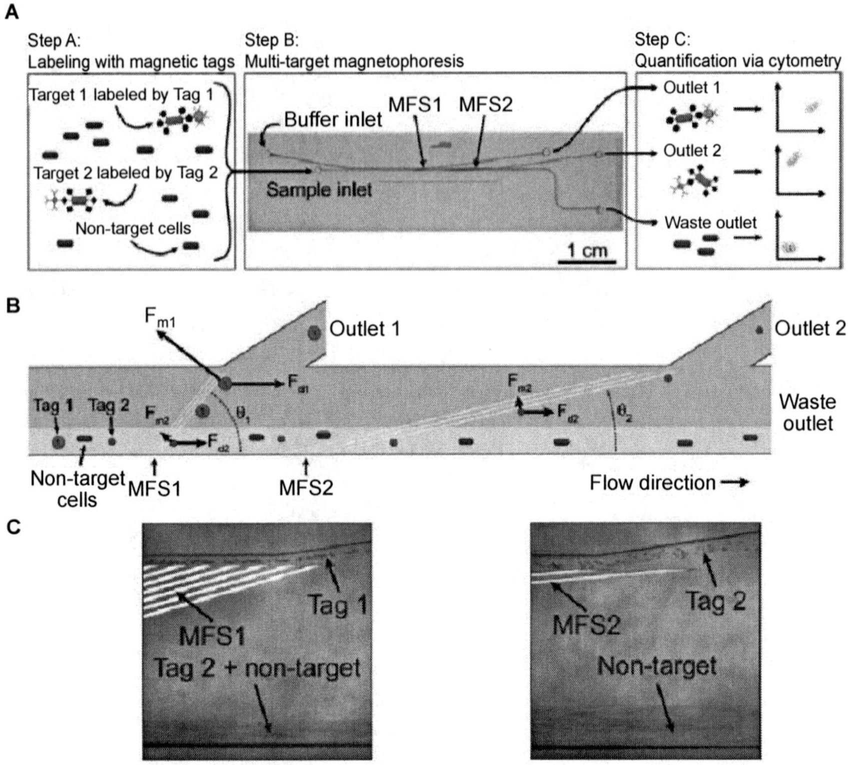

Figure 8.2: Schematic of the MT-MACS separation architecture.

Color image of this figure appears in the color plate section at the end of the book.

Moreover, many new methods for CTCs enrichment and detection combining immumomagnetic beads and microfludic device have been developed. For example, Superparamagnetic beads (SPMBs) in a flowing stream could be trapped in a regular hexagonal array of nickel micropillars integrated in a smart microfluidic device (Fig. 8.3) by a external magnetic field to form a magnetic bed (Liu et al. 2007). *In vivo* antibody modification to the trapped beads and subsequent A459 cells capture as well as washing steps were processed automatically by the controllable actuation of microvalves. Tumor cells were effectively enriched with 62%–74% efficiency while greatly reduced reagent consumption was gained though *in vivo* modification reaction. In another work reported by Lee's group (Lien et al. 2010), a 3D microfluidic incubator with antibody-coated magnetic beads in it was integrated into a microsystem. Over 90% targeted tumor cells were captured efficiently by a swirling effect generated by the incubator when blood samples passed through these beads, then enriched and purified by

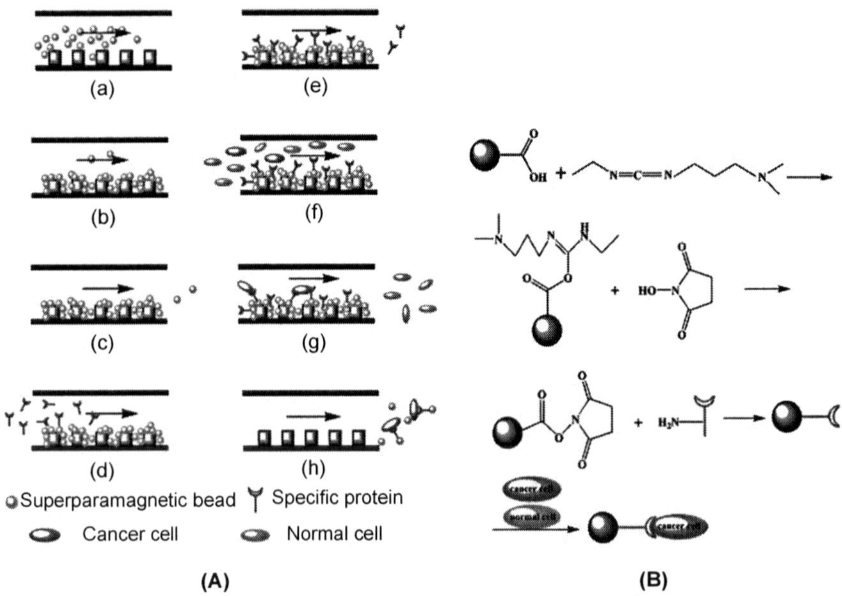

Figure 8.3: Schematic representation of the experiment protocol for cell capture and sorting.

an external magnet and a microfluidic control module. High sensitivity and rapid isolation made this device a potential automated platform for clinical use. More recently, a unique method vividly described as 'Ephesia' (Saliba et al. 2010) was proposed for tumor cells sorting with a capture yield better than 94% and the possibility of in situ cultivation (Fig. 8.4). In this process, columns of antibody coated magnetic beads self-assembled in a microfluidic channel by the interaction between ink dots patterned at the bottom of the channel and the beads in an external vertical magnetic field. The increased contact area of immumomagnetic beads and targeted cells facilitated the high capture efficiency along with reduced time and cost.

Besides these novel immumomagnetic bead methods combined with microfluidic system, Davis's group fabricated a robotic magnetic sweeper device (Talasaz et al. 2009) to enrich and separate circulating epithelial cells (CEpCs) or other rare cells with a high purity by 10^8-fold from the blood and easy accessibility that facilitates subsequent molecular analysis (Fig. 8.5). This MagSweeper used a magnetic rod covered with a plastic sheath to attract and separate immumomagnetic particles pre-bonded with MCF7 cells from the blood, followed by several rounds of capture-wash-release-recapture procedure to eliminate background cells, making itself an automated sweeper to capture MCF7 with an efficiency of 62% ± 7% and a purity of 51% ± 18%.

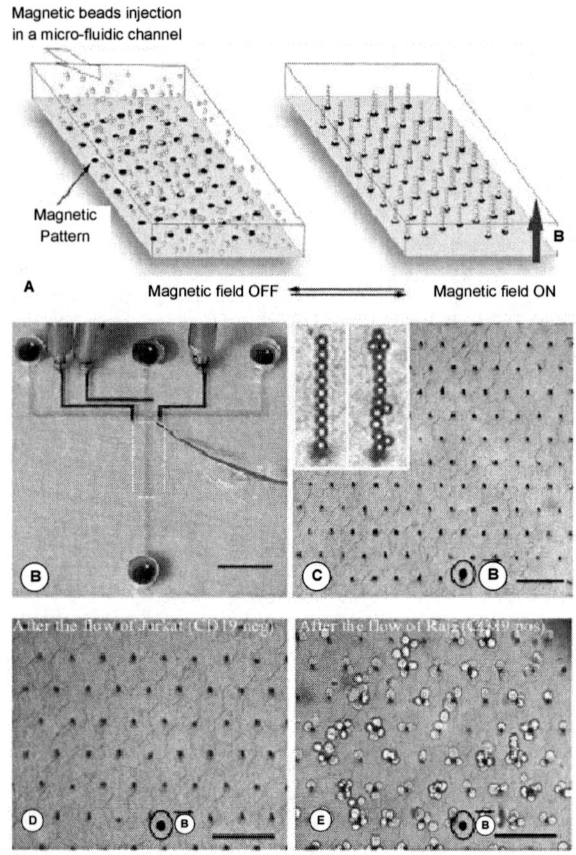

Figure 8.4: Principle and practical implementation of the 'Ephesia' System.

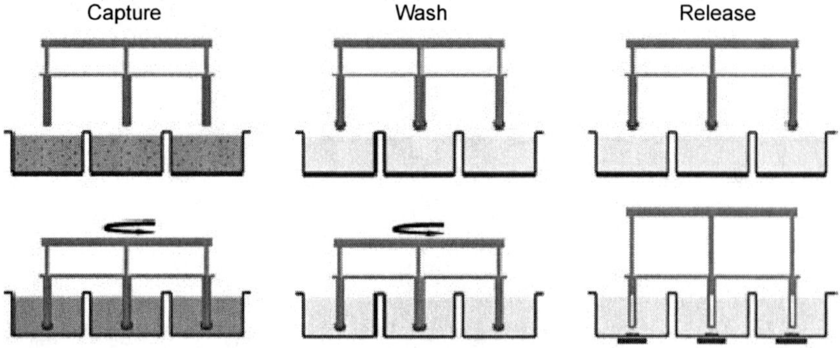

Figure 8.5: Schematic of the MagSweeper process.

Recently, another enlightening work using immumomagnetic selection was based on *in vivo* CTC detection. Zharov's group (Galanzha et al. 2009) used bio-functionalized magnetic nanoparticles combined with a two-color photoacoustic flow cytometry *in vivo* to enrich and detect the human breast cancer cell line MDA-MB-231 spiked in the bloodstream of mice. In this system two molecular agents were used, one bound to magnetic nanoparticles for the identification of the urokinase plasminogen activator receptors, while the other bound to golden carbon nanotubes targeting the folate receptors of cancer cells, thus improving the specificity and sensitivity of CTCs enrichment and the following photoacoustic detection.

Microfluidic-based Approaches

Based on the principles used in the enrichment procedures, microfluidic system based CTCs enrichment can be divided into two types, nonspecific size-based enrichment which separates cells just according to their sizes and specific molecular interaction based enrichment which separates cells by antigen-antibody identification or by aptamer conjugation.

SIZE-DEPENDANT MICROFLUIDIC CAPTURE

Kovic and Voldman (2007) reported a microfluidic cell sorting device combining microwell arrays with optical manipulation techniques to perform image-based cell sorting with easy manipulation and up to 155-fold enrichment. The microwells integrated at the bottom of the microchannel could detain cells one-for-one sedimentated in the flow with matched size. In another work, one step CTCs enrichment was achieved by using a microdevice with multiple arrays of crescent-shaped isolation wells integrated in the microchannel (Fig. 8.6) with at least 80% efficiency based on the physical differences such as stiffness and size between tumor cells and other blood constituents (Tan et al. 2009). Recently, a microfluidic device with dam structures (Fig. 8.6) in it was used to separate CTCs spiked into the blood (Li et al. 2010). A kind of lectin molecular with suitable concentration was added into the blood sample before the separation procedure to induce tumor cell agglomeration rather than blood constituents. The dimensions of the dam structure were essential to trap tumor cells and cell agglomerations in the gaps. Ninety nine percent purity with 84% efficiency and 94.4% viability made this device a feasible and useful platform for CTCs enrichment and the related studies.

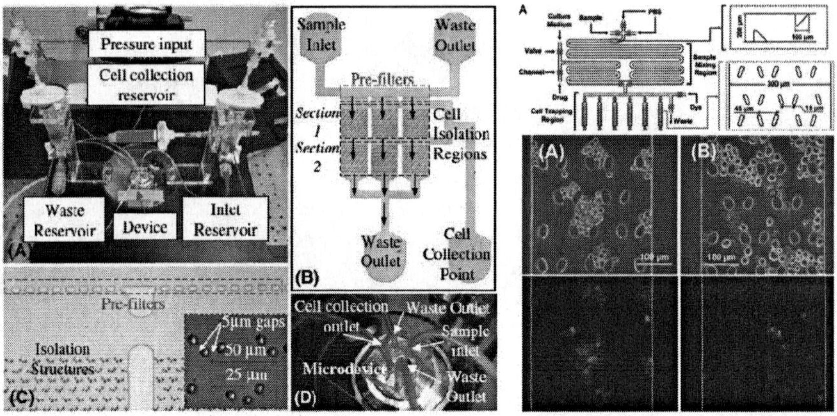

Figure 8.6: Microdevice with multiple arrays of crescent-shaped isolation wells integrated in the microchannel; microfluidic device with microdam structures.

Color image of this figure appears in the color plate section at the end of the book.

ANTIGEN-ANTIBODY INTERACTIONS

A polymer-based microfluidic device was fabricated and explored for CTCs capture in 2005 by Steven A. Soper's group (McCarley et al. 2005) as the difficulties of molecular modification and surface pattern on the microscale were solved by the generation of carboxylic acid group on PMMA and PC surface. Anti-EpCAM modified PMMA channel with narrow width (20 μm) could capture higher MCF7 compared to the wider one (50 μm). In another study reported by this group (Adams et al. 2008a), modified PMMA microfluidic with a series of high-aspect ration microchannels in it was used to isolate and enrich CTCs from whole blood of large volumes in a short period (<37 min) (Fig. 8.7). A high efficiency of 97% was obtained by the specific identification of MCF7 using mABs antibody coated on the walls of the channels and the increased contact time in these sinusoidal microwells.

The other representative microfluidic platform is known as 'CTC-chip' (Nagrath et al. 2007) which consists of an array of microposts modified with anti-EpCAM antibodies etched on silicon wafer (Fig. 8.8). The 'CTC-chip' was used to separate and enrich CTCs of five different cancers in whole blood procured from tumor patients with 99.1% sensitivity and 100% specificity, making it a potential tool in CTCs detection and clinical diagnosis. Following the 'CTC-chip', a 'HB-chip' (Stott et al. 2010), herringbone-chip with an array of herringbone grooves on the top of

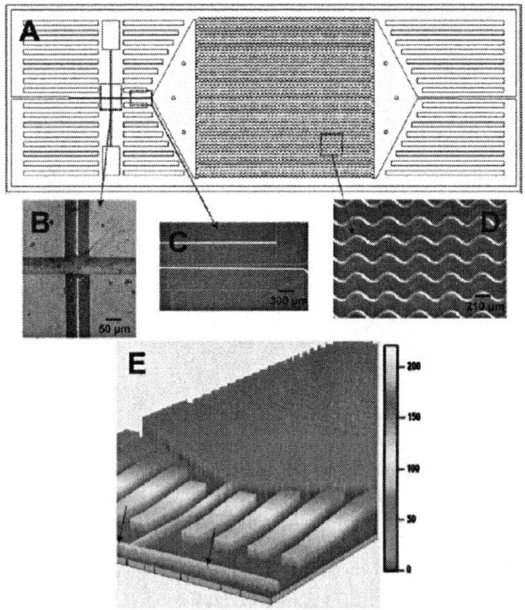

Figure 8.7: Modified PMMA microfluidic with a series of high-aspect ration microchannels in it was used to isolate and enrich CTCs from whole blood.

Color image of this figure appears in the color plate section at the end of the book.

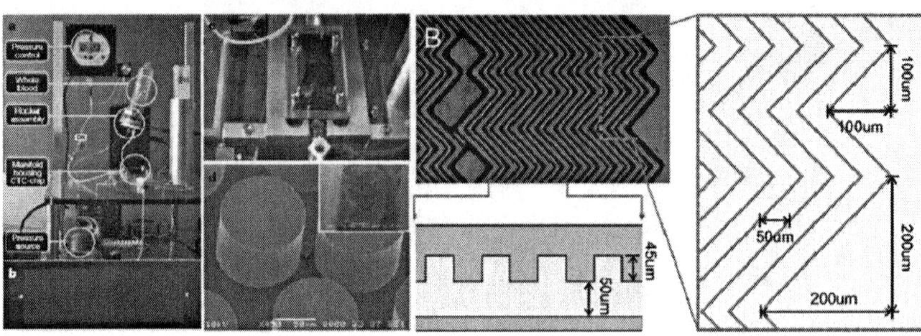

Figure 8.8: The 'CTC-chip' and the 'HB-chip'.

Color image of this figure appears in the color plate section at the end of the book.

microchannels was developed by this group (Fig. 8.8). These herringbone grooves could generate microvortices to mix blood cells and significantly increase the interactions between CTCs and the anti-body coated chip surface. Revealing CTCs microclusters by this chip, could be useful to metastasis diagnosis and is a result that deserves attention.

APTAMER IDENTIFICATION

Aptamers are single-strand nucleic acid oligomers targeting cell surface moleculars such as membrane proteins. They can be produced automatically and modified maturely to solid surfaces compared with antibodies. Recently, Tan's group (Phillips et al. 2009) designed an aptamer modified microfluidic device to capture target cells. A PDMS microchannel modified with aptamer sgc8 in it could specifically capture the T-cell Acute Lymphocytic Leukemia cells with 97% purity at an optimized flow situation. In another work (Xu et al. 2009), used the same principles to separate three types of cancer cells simultaneously using a developed microfluidic device, with a 135-fold enrichment and 96% purity (Fig. 8.9). The two devices used above have two features facilitating the detection efficiency, the transparency of PDMS allows monitoring and adjusting of separate parameters, the matched channel depth with cell size allows the increased interaction between cells and aptamers.

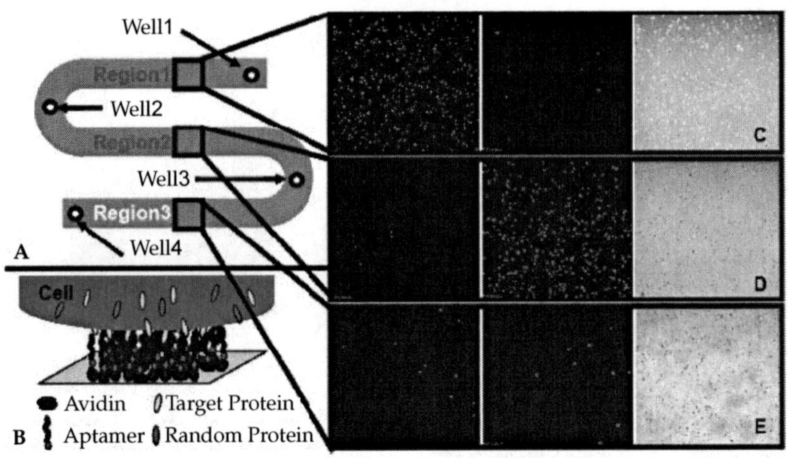

Figure 8.9: Aptamer-Based Microfluidic Device for Enrichment, Sorting, and Detection of Multiple Cancer Cells.

Color image of this figure appears in the color plate section at the end of the book.

In another work, anti prostate-specific membrane antigen (PSMA) aptamers were immobilized onto the inner walls of the microchannels created in the PMMA microfluidic device referred above for capturing LNCaP cells overexpressing PSMA from whole blood (Dharmasiri et al. 2009). The aptamer-coated HTMSU (high-throughput micro-sample unit) capture bed with channels which width was similar to the size of tumor cells can separate 90% of target cells, counted by an integrated conductivity sensor, from 1 mL input blood in ~29 min.

Nanosubstrate-based CTCs Capture

Considering the dimensions of the structures on the substrates involved in the CTCs enrichment methods mentioned above, they are all based on microscale. Recently, a new CTC capture platform (Fig. 8.10) was developed, using a three dimension (3D) nanostructured substrate coated with anti-EpCAM antibodies (Wang et al. 2009). An array of Si nanopillars (SiNPs) was etched on the surface of a silicon wafer. After a series of molecular modification processes, anti-EpCAM antibodies were grafted to the surface of the SiNPs. The prepared 'chips' were then used to capture MCF7 from whole blood, with a higher efficiency compared with the smooth Si wafers modified through the same procedure. Besides, the capture time and the lengths of SiNPs could influence the number of CTCs isolated. The author assumes that the enhanced local topographic interactions between 3D nanostructured substrates and cell surface nanostructures, in addition to antigen-antibody affinity, increase the cell capture efficiency. The convenient manipulation and cost-efficient chips makes themselves 'fly-papers' for CTC capture and a potential platform for clinical diagnosis.

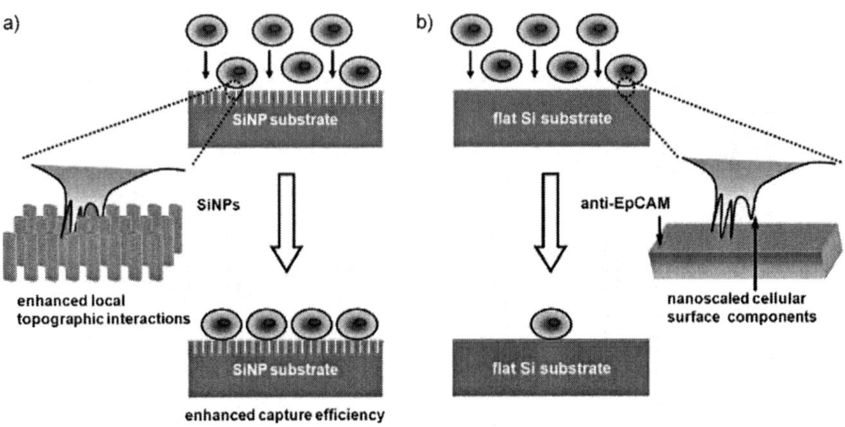

Figure 8.10: 3D nanostructure-based cell capture.

DISCUSSION AND PROSPECT

The detection of CTCs is very important for metastasis diagnosis, cancer prognosis, clinical personal treatment and the study of molecular biology. Recent research suggests that metastasis might even occur at the early stage of primary tumor progression, which increases the necessity of CTCs detection with high sensitivity and specificity. There is always an enrichment step before CTCs detection to enhance the sensitivity as CTCs are very few in human blood. Although many CTCs enrichment methods have been developed and tested experimentally, each with a certain superiority, the lack of a uniform experiment criterion makes it difficult to compare and assess them. This situation is not beneficial for the next clinical application and commercial exploitation. Thus it is essential to establish a series of relevant assessment systems to highlight the most valuable one. One factor that influences the specificity of CTCs enrichment is the lack of a specific maker, as not all CTCs express the surface target molecule at the same time while other healthy cells may abnormally express the target molecule simultaneously. Multiplex molecular identification may be a suitable solution.

Another proposal that deserves our attention is to combine the CTCs enrichment and detection together, in order to build a one-step, automatic and convenient CTCs analysis platform. The Cell Search System is a typical case using immunomagnetic beads for CTCs enrichment and a Cell-Spotter Analyzer for identifying CTCs after staining. However, it needs a semi-automated process. The microfluidic device may be a potential tool as it can stain, release and collect cells after the capture by automatically changing the reagent or washing solutions. Moreover, CTC identification can be integrated with microfluidics, such as the internal conductivity sensor and external microscopy.

Traditional CTCs detection and counting is based on fluorescence signal generated by digital fluorescence microscopy, fiber-optic array scanning technology (FAST) (Krivacic et al. 2004) or laser scanning cytometry (MAINTRAC®) (Rolle et al. 2005) after a series of cell fixation, penetration and staining procedures. The pre-treatment process is cumbersome and time-consuming. Besides, fluorescence reagents are toxic threatening the health of the operators and destroying the cells resulting in dead CTCs that cannot be used for the next biological study. New emerging electrical means can also be used to detect tumor cells. Steven A. Soper's group (Adams et al. 2008a, Dharmasiri et al. 2009) integrated a conductivity sensor in PMMA-based microfluidic devices to read CTCs released from the capture surface. The sensor detected tumor cells passing through the electrode, reading out solution conductivity based on specifically scaling factors with a near 100% efficiency. This process does not need cell staining and can isolate living cells for the next analysis. In another work, a PBA (phenylboronic acid)-modified gold electrode (Fig. 8.11) was used to detect sialic acid in the free state or expressed on the cell surface, based on the reversible and specific binding between PBA and SA, by utilizing a field effect transistor (FET) to monitor the charge density changes caused by the carboxyl groups of the SA (Matsumoto et al. 2009). This easy, label free process can be used to diagnosis tumor malignancy and metastasis (Matsumoto et al. 2010).

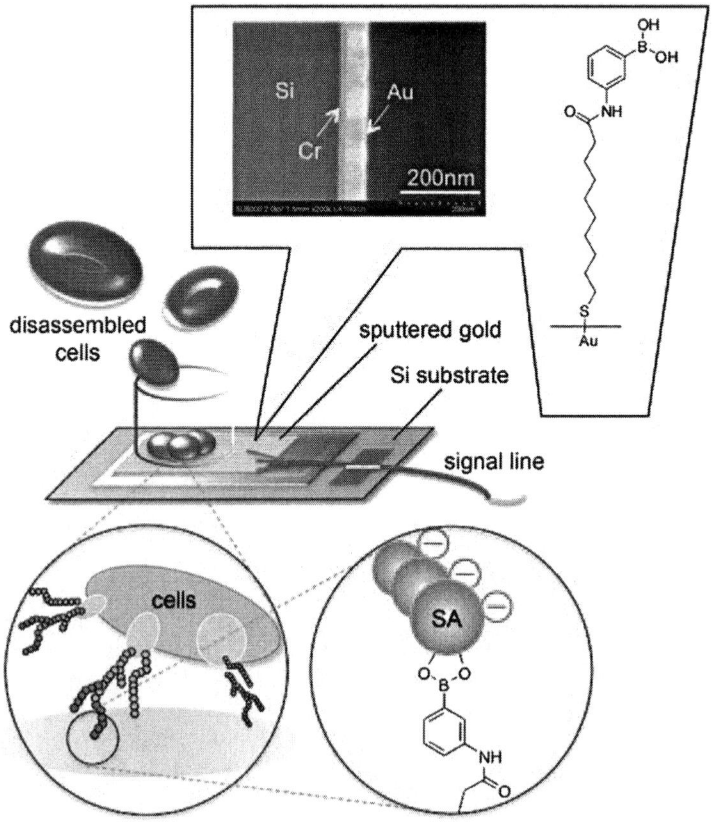

Figure 8.11: Sialic acid based electrochemical detection of tumor cells.
Color image of this figure appears in the color plate section at the end of the book.

Cell surface, as well as tumor cell surface (Domagala and Koss 1977, Ren et al. 1990), consists of nanostructures such as microvillis and filopodias. The *in vivo* environment that cells live in is also on the nanoscale such as extracellular matrix (ECMs). A lot of studies have been focusing on how nanoscale topography influences the behavior of cells (Chen et al. 1997, Curtis and Wilkinson 1997, Dalby et al. 2007, Flemming et al. 1999). Substrates with nanoscaled surface structures such as SiNWs have been used in anti-bioadhesion (Chen et al. 2009), cell isolation (Kim et al. 2010, Wang et al. 2009) and other fields. A recent study showed that silicon nanowires coated on the micobeads could enhance the adhesion between microbeads and Caco-2 cells which display a microvilliated structure similar to the structure *in vivo* (Fischer et al. 2009). A SiNWs based CTCs capture platform showed probable enhanced local topographic interactions between 3D nanostructured substrates and cell surface nanostructures (Wang et al. 2009). Although the mechanism behind this has not been clearly and exactly proved, this kind of nanostructures based CTC enrichment may be a promising choice for an applicative and low-cost platform.

CONCLUSION

Many methods for CTCs enrichment have been developed for the final CTCs detection and analysis. Among these methods, the traditional ones such as commercialized ISET and Cell Search System are widely used in most laboratorial and clinical studies. As nanotechnology emerges, some new enrichment devices have been designed with their own advantages aiming at a higher sensitivity and specificity, as well as convenience and low-cost. However, many problems have not been solved such as the lack of a specific tumor marker. In the future a systemic assessment procedure needs to be set up, an excellent integrative CTCs enrichment-detection platform can be built. More effort should be paid to these two main issues.

REFERENCES

Adams, A.A. and P.I. Okagbare, J. Feng, M.L. Hupert, D. Patterson, J. Gottert, R.L. McCarley, D. Nikitopoulos, M.C. Murphy and S.A. Soper. 2008a. Highly efficient circulating tumor cell isolation from whole blood and label-free enumeration using polymer-based microfluidics with an integrated conductivity sensor. J Am Chem Soc 130: 8633–8641.

Adams, J.D. and U. Kim and H.T. Soh. 2008b. Multitarget magnetic activated cell sorter. P Natl Acad Sci USA 105: 18165–18170.

Ashworth, T.R. 1869. A case of cancer in which cells similar to those in the tumors were seen in the blood after death. Aust Med J 14.

Balic, M., N. Dandachi, G. Hofmann, H. Samonigg, H. Loibner, A. Obwaller, A. van der Kooi, A.G.J. Tibbe, G.V. Doyle, L.W.M.M. Terstappen et al. 2005. Comparison of two methods for enumerating circulating tumor cells in carcinoma patients. Cytom Part B-Clin Cy 68B, 25–30.

Bilkenroth, U. and H. Taubert, D. Riemann, U. Rebmann, H. Heynemann and A. Meye. 2001. Detection and enrichment of disseminated renal carcinoma cells from peripheral blood by immunomagnetic cell separation. Int J Cancer 92: 577–582.

Chen, C.S. and M. Mrksich, S. Huang, G.M. Whitesides and D.E. Ingber. 1997. Geometric control of cell life and death. Science 276: 1425–1428.

Chen, L. and M.J. Liu, H. Bai, P.P. Chen, F. Xia, D. Han and L. Jiang. 2009. Antiplatelet and Thermally Responsive Poly(N-isopropylacrylamide) Surface with Nanoscale Topography. J Am Chem Soc 131: 10467–10472.

Curtis, A. and C. Wilkinson. 1997. Topographical control of cells. Biomaterials 18: 1573–1583.

Dalby, M.J. and N. Gadegaard, R. Tare, A. Andar, M.O. Riehle, P. Herzyk, C.D.W. Wilkinson and R.O.C.Oreffo. 2007. The control of human mesenchymal cell differentiation using nanoscale symmetry and disorder. Nat Mater 6: 997–1003.

Datta, Y.H. and P.T. Adams, W.R. Drobyski, S.P. Ethier, V.H. Terry and M.S. Roth. 1994. Sensitive Detection of Occult Breast-Cancer by the Reverse-Transcriptase Polymerase Chain-Reaction. J Clin Oncol 12: 475–482.

Dharmasiri, U. and S. Balamurugan, A.A. Adams, P.I. Okagbare, A. Obubuafo and S.A. Soper. 2009. Highly efficient capture and enumeration of low abundance prostate cancer cells using prostate-specific membrane antigen aptamers immobilized to a polymeric microfluidic device. Electrophoresis 30: 3289–3300.

Dittrich, P.S. and A. Manz. 2006. Lab-on-a-chip: microfluidics in drug discovery. Nat Rev Drug Discov 5: 210–218.

Domagala, W. and L.G. Koss. 1977. Configuration of Surfaces of Human Cancer-Cells in Effusions—Scanning Electron-Microscopic Study of Microvilli. Virchows Arch B 26: 27–42.

Fischer, K.E. and B.J. Aleman, S.L. Tao, R.H. Daniels, E.M. Li, M.D. Bunger, G. Nagaraj, P. Singh, A. Zettl and T.A. Desai. 2009. Biomimetic Nanowire Coatings for Next Generation Adhesive Drug Delivery Systems. Nano Lett 9: 716–720.

Flatmark, K., K. Bjornland, H.O. Johannessen, E. Hegstad, R. Rosales, L. Harklau, J.H. Solhaug, R.S. Faye, O. Soreide, O. Fodstad et al. 2002. Immunomagnetic detection of micrometastatic cells in bone marrow of colorectal cancer patients. Clin Cancer Res 8: 444–449.

Flemming, R.G. and C.J. Murphy, G.A. Abrams, S.L. Goodman and P.F. Nealey. 1999. Effects of synthetic micro- and nano-structured surfaces on cell behavior. Biomaterials 20: 573–588.

Galanzha, E.I. and E.V. Shashkov, T. Kelly, J.W. Kim, L.L. Yang and V.P. Zharov. 2009. *In vivo* magnetic enrichment and multiplex photoacoustic detection of circulating tumour cells. Nat Nanotechnol 4: 855–860.

Gijs, M.A.M. and F. Lacharme and U. Lehmann. 2010. Microfluidic Applications of Magnetic Particles for Biological Analysis and Catalysis. Chemical Reviews 110: 1518–1563.

Goeminne, J.C. and T. Guillaume and M. Symann. 2000. Pitfalls in the detection of disseminated non-hematological tumor cells. Ann Oncol 11: 785–792.

Hong, J.W. and S.R. Quake. 2003. Integrated nanoliter systems. Nat Biotechnol 21: 1179–1183.

Kim, S.T. and D.J. Kim, T.J. Kim, D.W. Seo, T.H. Kim, S.Y. Lee, K. Kim, K.M. Lee, and S.K. Lee. 2010. Novel Streptavidin-Functionalized Silicon Nanowire Arrays for CD4(+) T Lymphocyte Separation. Nano Lett 10: 2877–2883.

Kovac, J.R. and J. Voldman. 2007. Intuitive, image-based cell sorting using optofluidic cell sorting. Anal Chem 79: 9321–9330.

Krivacic, R.T., A. Ladanyi, D.N. Curry, H.B. Hsieh, P. Kuhn, D.E. Bergsrud, J.F. Kepros, T. Barbera, M.Y. Ho, L.B. Chen et al. 2004. A rare-cell detector for cancer. P Natl Acad Sci USA 101: 10501–10504.

Li, L. and W.M. Liu, J.C. Wang, Q. Tu, R. Liu and J.Y. Wang. 2010. Lectin-aided separation of circulating tumor cells and assay of their response to an anticancer drug in an integrated microfluidic device. Electrophoresis 31: 3159–3166.

Lien, K.Y. and Y.H. Chuang, L.Y. Hung, K.F. Hsu, W.W. Lai, C.L. Ho, C.Y. Chou and G.B. Lee. 2010. Rapid isolation and detection of cancer cells by utilizing integrated microfluidic systems. Lab Chip 10: 2875–2886.

Liu, Y.J. and S.S. Guo, Z.L. Zhang, W.H. Huang, D. Baigl, M. Xie, Y. Chen and D.W.Pang. 2007. A micropillar-integrated smart microfluidic device for specific capture and sorting of cells. Electrophoresis 28: 4713–4722.

Martin, V.M. and C. Siewert, A. Scharl, T. Harms, R. Heinze, S. Ohl, A. Radbruch, S.Miltenyi and J. Schmitz. 1998. Immunomagnetic enrichment of disseminated epithelial tumor cells from peripheral blood by MACS. Exp Hematol 26: 252–264.

Matsumoto, A. and N. Sato, K. Kataoka and Y. Miyahara. 2009. Noninvasive Sialic Acid Detection at Cell Membrane by Using Phenylboronic Acid Modified Self-Assembled Monolayer Gold Electrode. J Am Chem Soc 131: 12022–12033.

Matsumoto, A. and H. Cabral, N. Sato, K. Kataoka and Y. Miyahara. 2010. Assessment of Tumor Metastasis by the Direct Determination of Cell-Membrane Sialic Acid Expression. Angewandte Chemie-International Edition 49: 5494–5497.

McCarley, R.L. and B. Vaidya, S.Y. Wei, A.F. Smith, A.B. Patel, J. Feng, M.C. Murphy and S.A. Soper. 2005. Resist-free patterning of surface architectures in polymer-based microanalytical devices. J Am Chem Soc 127: 842–843.

Müller, V. and N. Stahmann, S. Riethdorf, T. Rau, T. Zabel, A. Goetz, F. Jänicke and K. Pantel. 2005. Circulating Tumor Cells in Breast Cancer: Correlation to Bone Marrow Micrometastases, Heterogeneous Response to Systemic Therapy and Low Proliferative Activity. Clin Cancer Res 11: 3678–3685.

Nagrath, S., L.V. Sequist, S. Maheswaran, D.W. Bell, D. Irimia, L. Ulkus, M.R. Smith, E.L. Kwak, S. Digumarthy, A. Muzikansky et al. 2007. Isolation of rare circulating tumour cells in cancer patients by microchip technology. Nature 450: 1235–U1210.

Naume, B. and E. Borgen, K. Beiske, T.K. Herstad, G. Ravnas, A. Renolen, S.Trachsel, K.ThraneSteen, S.Funderud and G.Kvalheim. 1997. Immunomagnetic techniques for the enrichment and detection of isolated breast carcinoma cells in bone marrow and peripheral blood. J Hematother 6: 103–114.

Naume, B. and E. Borgen, J.M. Nesland, K. Beiske, E. Gilen, A. Renolen, G. Ravnas, H. Qvist, R. Karesen and G.Kvalheim. 1998. Increased sensitivity for detection of micrometastases in bone-marrow/peripheral-blood stem-cell products from breast-cancer patients by negative immunomagnetic separation. Int J Cancer 78: 556–560.

Ntouroupi, T.G., S.Q.Ashraf, S.B. McGregor, B.W. Turney, A. Seppo, Y. Kim, X. Wang, M.W. Kilpatrick, P. Tsipouras, T. Tafas et al. 2008. Detection of circulating tumour cells in peripheral blood with an automated scanning fluorescence microscope. Brit J Cancer 99: 789–795.

Paget, S. 1889. The Distribution of Secondary Growths in Cancer of the Breast. The Lancet 133: 571–573.

Pantel, K. and R.J. Cote and O. Fodstad. 1999. Detection and clinical importance of micrometastatic disease. J Natl Cancer I 91: 1113–1124.

Phillips, J.A. and Y. Xu, Z. Xia, Z.H. Fan and W.H. Tan. 2009. Enrichment of Cancer Cells Using Aptamers Immobilized on a Microfluidic Channel. Analytical Chemistry 81: 1033–1039.

Pinzani, P. and B. Salvadori, L. Simi, S. Bianchi, V. Distante, L. Cataliotti, M. Pazzagli and C. Orlando. 2006. Isolation by size of epithelial tumor cells in peripheral blood of patients with breast cancer: correlation with real-time reverse transcriptase-polymerase chain reaction results and feasibility of molecular analysis by laser microdissection. Hum Pathol 37: 711–718.

Qu, B.Y. and Z.Y. Wu, F. Fang, Z.M. Bai, D.Z. Yang and S.K. Xu. 2008. A glass microfluidic chip for continuous blood cell sorting by a magnetic gradient without labeling. Anal Bioanal Chem 392: 1317–1324.

Racila, E. and D. Euhus, A.J. Weiss, C. Rao, J. McConnell, L.W.M.M. Terstappen and J.W. Uhr. 1998. Detection and characterization of carcinoma cells in the blood. P Natl Acad Sci USA 95: 4589–4594.

Ren, J. and J. Hamada, F. Okada, N. Takeichi, K. Morikawa, M. Hosokawa and H. Kobayashi. 1990. Correlation between the Presence of Microvilli and the Growth or Metastatic Potential of Tumor-Cells. Jpn J Cancer Res 81: 920–926.

Riethdorf, S., H. Fritsche, V. Muller, T. Rau, C. Schindibeck, B. Rack, W. Janni, C. Coith, K. Beck, F. Janicke et al. 2007. Detection of circulating tumor cells in peripheral blood of patients with metastatic breast cancer: A validation study of the Cell Search system. Clin Cancer Res 13: 920–928.

Rolle, A. and R. Gunzel, U. Pachmann, B. Willen, K. Hoffken and K. Pachmann. 2005. Increase in number of circulating disseminated epithelial cells after surgery for non-small cell lung cancer monitored by MAINTRAC(R) is a predictor for relapse: A preliminary report. World Journal of Surgical Oncology 3: 18–18.

Rosi, N.L. and C.A. Mirkin. 2005. Nanostructures in biodiagnostics. Chemical Reviews 105: 1547–1562.

Saliba, A.E., L. Saias, E. Psychari, N. Minc, D. Simon, F.C. Bidard, C. Mathiot, J.Y. Pierga, V. Fraisier, J. Salamero et al. 2010. Microfluidic sorting and multimodal typing of cancer cells in self-assembled magnetic arrays. Proceedings of the National Academy of Sciences of the United States of America 107: 14524–14529.

Schoenfeld, A. and K.H. Kruger, J. Gomm, H.D. Sinnett, J.C. Gazet, N. Sacks, H.G. Bender, Y. Luqmani and R.C. Coombes. 1997. The detection of micrometastases in the peripheral blood and bone marrow of patients with breast cancer using immunohistochemistry and reverse transcriptase polymerase chain reaction for keratin 19. Eur J Cancer 33: 854–861.

Smith, B.M. and M.J. Slade, J. English, H. Graham, M. Luchtenborg, H.D. Sinnett, N.C.P. Cross and R.C. Coombes. 2000. Response of circulating tumor cells to systemic therapy in patients with metastatic breast cancer: Comparison of quantitative polymerase chain reaction and immunocytochemical techniques. J Clin Oncol 18: 1432–1439.

Steeg, P.S. 2006. Tumor metastasis: mechanistic insights and clinical challenges. Nat Med 12: 895–904.

Stott, S.L., C.H. Hsu, D.I. Tsukrov, M. Yu, D.T. Miyamoto, B.A. Waltman, S.M. Rothenberg, A.M. Shah, M.E. Smas, G.K. Korir et al. 2010. Isolation of circulating tumor cells using a microvortex-generating herringbone-chip. P Natl Acad Sci USA 107: 18392–18397.

Talasaz, A.H., A.A. Powell, D.E. Huber, J.G. Berbee, K.H. Roh, W. Yu, W.Z. Xiao, M.M.Davis, R.F. Pease, M.N. Mindrinos et al. 2009. Isolating highly enriched populations of circulating epithelial cells and other rare cells from blood using a magnetic sweeper device. P Natl Acad Sci USA 106: 3970–3975.

Tan, S.J. and L. Yobas, G.Y.H. Lee, C.N. Ong and C.T. Lim. 2009. Microdevice for the isolation and enumeration of cancer cells from blood. Biomed Microdevices 11: 883–892.

Vona, G., A. Sabile, M. Louha, V. Sitruk, S. Romana, K. Schutze, F. Capron, D. Franco, M. Pazzagli, M. Vekemans et al. 2000. Isolation by size of epithelial tumor cells—A new method for the immunomorphological and molecular characterization of circulating tumor cells. American Journal of Pathology 156: 57–63.

Wang, S.T., H. Wang, J. Jiao, K.J. Chen, G.E. Owens, K.I. Kamei, J. Sun, D.J. Sherman, C.P. Behrenbruch, H. Wu et al. 2009. Three-Dimensional Nanostructured Substrates toward Efficient Capture of Circulating Tumor Cells. Angewandte Chemie-International Edition 48: 8970–8973.

Xu, Y. and J.A. Phillips, J.L. Yan, Q.G. Li, Z.H. Fan and W.H. Tan. 2009. Aptamer-Based Microfluidic Device for Enrichment, Sorting, and Detection of Multiple Cancer Cells. Analytical Chemistry 81: 7436–7442.

Yager, P. and T. Edwards, E. Fu, K. Helton, K. Nelson, M.R. Tam and B.H. Weigl. 2006. Microfluidic diagnostic technologies for global public health. Nature 442: 412–418.

Zheng, S. and H. Lin, J.Q. Liu, M. Balic, R. Datar, R.J. Cote and Y.C. Tai. 2007. Membrane microfilter device for selective capture, electrolysis and genomic analysis of human circulating tumor cells. J Chromatogr A 1162: 154–161.

9

Detection of Bacterial Habits: Single Planktonic Cells and Assembled Biofilm

Akira Ishihama

INTRODUCTION

In the early stages of molecular biology, bacteria such as the Gram-negative *Escherichia coli* and the Gram-positive *Bacillus subtilis* were used as model organisms of molecular biology relying on the belief that cultures of single bacterial species are homogenous in cell populations. Recently, however, this concept has been challenged and our up-to-date understanding is that in natural environments, bacteria form organized communities, in which each cell plays a unique and specific role different from other cells. In exponential growth phase under laboratory culture conditions suitable for bacterial growth such as in rich media and under enough aeration, bacteria grow as planktonic single cell states. Most of the important knowledge in molecular biology such as the mechanisms and regulation of gene expression were established based on the experimental data obtained using model bacteria growing under laboratory culture conditions. Under stressful conditions in nature and even in controlled culture conditions such as in poor media and under anaerobic conditions, however, bacteria gather together into sessile communities, attached on solid surfaces, forming biofilms. Nowadays, it is widely accepted that biofilm is the major habit of microorganisms in nature.

Hosei University, Department of Frontier Bioscience, Koganei, Tokyo 184-8584, Japan; E-mail: aishiham@hosei.ac.jp

Gram-negative *Escherichia coli*, the best-studied model prokaryote, contains about 4,500 genes on its genome (Riley et al. 2006), but under laboratory culture conditions, only one-fourth of the genes are expressed and others remain silent. About one third of the gene functions remain unidentified, which are believed to be expressed under stress conditions and are needed for bacterial survival in nature. The bacterial response to stressful environments in nature involves various combinations of genes (for reviews see Ishihama 2000, 2010). At present, however, little is known about the molecular events leading to the formation of biofilm, and the molecular characteristics of various cell populations within a single and the same biofilm. Even now most of the research in molecular genetic field is being performed using the whole set of cultured bacterial cells, thereby leading to yield the average values of various differentiated cells. To make a breakthrough of this serious problem, a novel experimental system(s) is needed for monitoring the events carried out in individual cells within a single and the same culture. This chapter focuses on our recent efforts for development of the detection systems of cellular differentiation pathway of the model prokaryote, *Escherichia coli*, under laboratory culture conditions and the experimental systems for the identification of differentiated individual cells.

MATERIALS AND METHODS

1. Genomic SELEX Screening

For identification of regulation targets of DNA-binding transcription factors (TFs), we developed and employed the improved Genomic SELEX system. In brief, genome DNA of *E. coli* W3110 was sonicated to generate fragments of 200–300 bp in length (Shimada et al. 2005). The *E. coli* DNA library was constructed after cloning of these 100-300 bp DNA fragments into plasmid pBR322 at *Eco*RV site. In each experiment, a collection of these 100–300 bp DNA fragments could be regenerated by PCR-amplification using the *E. coli* DNA library plasmids as templates. For the genomic SELEX screening of TF-binding sequences, a mixture of DNA fragments and purified His-tagged TF was incubated for 30 min at 37°C. The mixture was directly applied onto Ni-NTA column, and DNA-TF complexes bound to the column were eluted with an imidazole-containing elution buffer. DNA sequences of DNA fragments isolated by Genomic SELEX screening were determined by two procedures, SELEX-clos (cloning and sequencing) and SELEX-chip (DNA chip analysis).

For SELEX-clos analysis, TF-bound DNA fragments were PCR amplified and PCR products were cloned into pT7 Blue-T vector (Novgen) and transformed into *E. coli* DH5α. After transformation, plasmids were recovered from as many independent transformants as possible, and subjected to DNA sequencing (Shimada et al. 2005). For SELEX-chip analysis, SELEX DNA fragments were labeled with Cy5 while the original mixture of genome DNA fragments was labeled with Cy3. Mixtures of Cy5- and Cy3-labeled DNA were subjected to hybridization with DNA tilling-array. The array consists of 43,450 species of 60 b-long DNA probe, which are designed to cover the entire *E. coli* genome at 105 bp interval (Oxford Gene Technology, Oxford, UK). The Cy5/Cy3 ratio was measured for each peak, and plotted against the corresponding position on the *E. coli* genome (Teramoto et al. 2000b, Shimada et al. 2011a and 2011b).

2. Percoll Density Gradient Centrifugation for Cell Separation

Percoll gradient centrifugation of *E. coli* cultures was carried out as described previously (Makinoshima et al. 2002). In brief, the cells were harvested by centrifugation and resuspended in appropriate volumes of phosphate-buffered saline (PBS). A 0.5 ml cell suspension was layered, together with a set of colored density-markers, on top of 9.5 ml of the preformed Percoll gradient, which was prepared by centrifugation of 80% Percoll for 1 hr at 20,000 rpm at 4°C in a Beckman SW40.1Ti rotor. After centrifugation for 45 min at 20,000 rpm at 4°C, the pattern of cell distribution along Percoll gradient was recorded with a Nikon digital camera and then the Percoll gradient was fractionated using a Piston gradient fractionator (Biocomp).

3. Construction of Cell Chips for Single Cell Observation

The transparent silicone elastomer, poly(dimethylsiloxane) (PDMS), was used for construction of the cell chip. The transparent microelectrodes of ITO (indium tin oxide) were patterned by photolithography on the glass plate, and the plate was bonded to position the microelectrodes in the micro-channel (Arai et al. 2003, Teramoto et al. 2010a). Wild-type *E. coli* K-12 cells carrying a promoter assay vector were cultured in LB medium at 30°C, into which various concentrations of poly-(*N*-isopropylacrylamide) (PNIPAAm) were added at 25°C. The cell suspension was passed into the cell chip at room temperature (25°C) through the input micro-channel using a controlled Peristaltic pump. Upon raising the local temperature to 30°C around the ITO electrode, a portion of the culture was converted into PNIPAAm gel (Ichikawa et al. 2005, Teramoto et al. 2010a). After gelation of portions of bacterial culture, fresh media can be circulated through the same channel in the case of one-channel cell chip. For addition of effectors affecting the promoter activity, the two-channel chip was used, into which the effectors were pumped using the second input channel. Gelation of PNIPPAAm solution is concentration-dependent and temperature-sensitive. Thus for stable maintenance of fixed bacteria within gels for prolonged time periods, the temperature sensor was patterned by photolithograny near the ITO electrode on the glass plate.

4. Measurement of the Promoter Activity and Regulation

The promoter strength and regulation was measured by using two types of the promoter assay vector using either fluorescent proteins (GFP and RFP) or *lacZ*-coded beta-galactosidase as reporters. The vector pTFP (two-fluorescent protein vector) or pGRP (GFP and RFP vector) carries two types of fluorescent protein genes, one for red fluorescent protein RFP and the other for green fluorescent protein GFP as reporters. The construction of promoter assay vectors was performed according to the original procedure (Shimada et al. 2004). The GFP gene was aligned as to be under the control of a test promoter while the RFP gene was inserted to be under the control of a reference promoter *lac*UV5. The test promoter sequences upstream from the respective translation initiation codon up to about 500 bp were PCR-amplified and inserted into pGRP. The promoter assay vectors thus constructed were transformed into wild-type and mutants, each lacking one of the transcription factor genes. For measurement of the fluorescent intensity of RFP or GRP expressed in *E. coli*, cells grown in LB or M9-0.4% glucose media for various times were harvested by centrifugation, resuspended in PBS(-) to give approximately the same cell density of 0.6 A600nm for all the samples. For measurement of bulk fluorescence, aliquots of 0.3 ml cell suspension were applied into 0.4-ml flat-bottom 96 wells, and the fluorescence was measured with a FL600 Bio-Tek microplate reader (Bio-Tek Instruments, USA) or a Wallac 1420 ARVOsx (Perkin-Elmer Life Sciences, USA). GFP was measured using 485-nm excitation and 535-nm emission and RFP was measured using 544-nm excitation and 590-nm emission. The net value of fluorescence was measured after subtraction of the background fluorescence determined using *E. coli* cultures with pDFP vector without promoter insertions. The fluorescence intensity of GFP by the test promoter was normalized using the equation (X/Y)/(A/B), in which X and Y indicate the fluorescence intensities of GFP (test promoter) and RFP (*lacUV5* promoter), respectively, while A and B represent the fluorescence intensities of GFP (*lacUV5* promoter) and RFP (*lacUV5* promoter), respectively.

For the construction of *lacZ* reporter vectors, DNA fragments, each containing 200- to 500 bp-sequence of test promoters, were prepared by PCR using *E. coli* KP7600 genome DNA as a template, and inserted into pRS551 (Simons et al. 1987). The test promoter-*lacZ* fusion plasmid was transformed into *E. coli* MC4100 (Casadaban, 1976) and the transformant was used as the host for preparation of λRS45. The recombinant phage containing promoter-*lacZ* fusion was isolated from the resulting phage lysate and infected onto recipient *E. coli* for construction of single-copy promoter-*lacZ* fusion strains. Transformants were grown in YESCA medium and subjected to measuring β-galactosidase activity in the presence of o-nitrophenyl-D-galactopyranoside (Miller 1972).

5. Detection of Transcription Factor-DNA Interactions

For detection of DNA-binding activity of test transcription factors, gel shift and DNase I foot-printing assays were employed. Fluorescent-labeled promoter probes were generated by PCR amplification using a pair of primers, promoter plasmids as templates, and Ex Taq DNA polymerase (Ogasawara et al. 2007a and 2007b). The DNase-I footprinting assay was carried out essentially under the standard reaction conditions (Ogasawara et al. 2007a and 2007b). In brief, an FITC-labeled probe was incubated at 37°C for 30 min with purified TF, and immediately subjected to DNA digestion by adding an aliquot of DNase I. After digestion for 30 s at 25°C, the reaction was terminated by the addition of phenol. DNA was precipitated from aqueous layer by ethanol, dissolved in formamide dye solution, and analyzed by electrophoresis in the presence of 8 M urea.

For primer extension analysis, RNA was purified according to the standard hot-phenol extraction method (Yamamoto and Ishihama 2003, Ogasawara et al. 2007a and 2007b). The primer extension reaction was initiated by the addition of a small amount of a reverse transcriptase. After incubation for 1 hr at 50°C, DNA was extracted with phenol, precipitated with ethanol, and subjected to electrophoresis on a 6% polyacrylamide sequencing gel containing 8 M urea using DSQ-500L (Shimadzu) for direct detection of fluorescence.

RESULTS AND DISCUSSION

1. Detection of Regulatory Roles of Transcription Factors for Growth Phase-Coupled Regulation

1.1 Promoter Selectivity Control of RNA Polymerase

After the genome sequencing, the research focus of bacterial molecular genetics shifted from the regulation of individual genes toward understanding the global regulation of genome transcription. For instance, *E. coli* contains a total of 4,450 genes (Riley et al. 2006) that are organized in about 2,000–3,000 operons, each under the control of a specific promoter. For transcription of stress-response operons, however, multiple promoters are often involved, each responding to a different environmental condition and each being recognized by a different sigma factor. The intracellular concentrations of seven species of the sigma factor in *E. coli* vary depending on the growth phase and growth conditions (Jishage and Ishihama 1995, Jishage et al. 1996). On the basis of the binding affinity of each sigma factor to the core RNA polymerase and the intracellular concentration of each sigma factor, we can predict the level of seven species of the holoenzyme, each recognizing a specific set of promoters. In addition, the promoter selectivity of RNA polymerase holoenzyme is further modulated by interaction with transcription factors (Fig. 9.1).

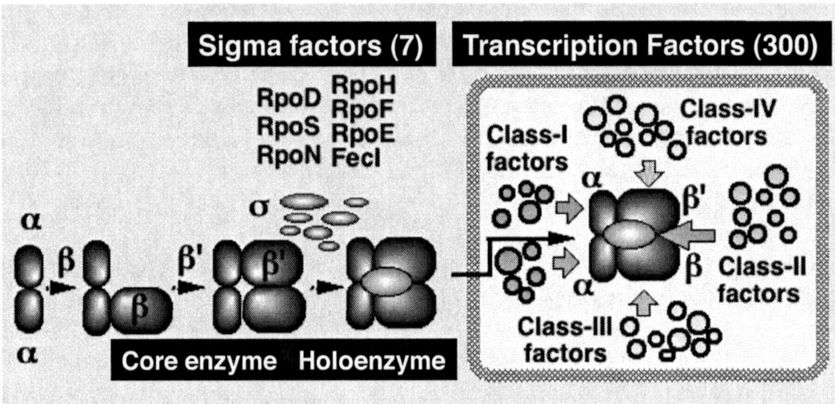

Figure 9.1: Functional differentiation of RNA polymerase. RNA polymerase core enzyme with the catalytic activity of RNA polymerization is assembled under a sequential manner in the order: $2\alpha > \alpha_2 > \alpha_2\beta > \alpha_2\beta\beta'$ (Ishihama 1990). The core enzyme is converted into holoenzyme after binding one of seven species of sigma subunit, the promoter recognition subunit. The promoter recognition specificity is further modulated after interaction with transcription factor(s). A total of about 300 species of transcription factors exist in *E. coli*, each interacting one of four holoenzyme subunits (Ishihama 2009, 2010).

A total of about 300 species of transcription factors, mostly with DNA-binding activity, exist in *E. coli*, each playing a specific regulatory role in transcription control of a set of genes (Ishihama 2009, 2010). Even for the best characterized model prokaryote *E. coli*, the regulatory role has not been identified for about one third of the transcription factors supposedly because these hitherto uncharacterized transcription factors are not expressed under laboratory culture conditions and may be needed for transcription of the genes expressed only under stress conditions in nature. In order to get insights into the regulation of genome expression, it is absolutely necessary to identify the regulatory functions of all 300 transcription factors in *E. coli*.

1.2 Genomic SELEX Screening for Regulation Targets by Transcription Factors

Our current effort focuses on the identification of whole sets of regulation target promoters, genes and operons under the control of each transcription factor. For this purpose, we have developed the Genomic SELEX screening system, in which DNA fragments bound by a test transcription factor are isolated from a mixture of genome DNA fragments (Fig. 9.2). After sequence analysis of these DNA segments by SELEX-clos (Shimada et al. 2005) or mapping on the *E. coli* genome using tilling array by SELEX-chip (Shimada et al. 2007), regulation targets of the test transcription factor can be predicted. So far we have performed Genomic SELEX screening for more than 200 species of *E. coli* transcription factors. The Genomic SELEX screening is, in particular, useful for identification of regulation targets by as yet uncharacterized transcription factors, because estimation of their regulatory functions is difficult without the knowledge of environmental conditions where these regulators are needed. For instance, we identified a total of 688 binding targets for hitherto uncharacterized YgiP and based on the distribution knowledge of its binding sites along the *E. coli* genome, we proposed that YgiP is a novel nucleoid protein Dan under anaerobic growth conditions [and thus renamed Dan] (Teramoto et al. 2010b).

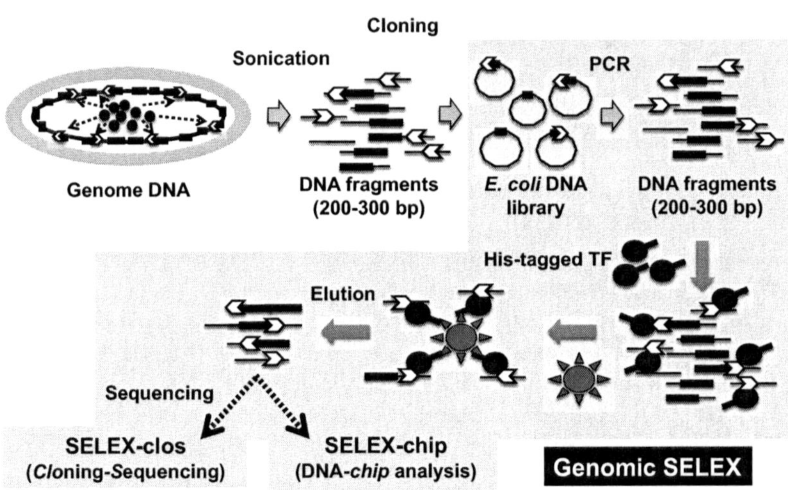

Figure 9.2: Genomic SELEX screening of regulation targets by transcription factors. Regulation targets of transcription factors can be identified after identification of recognition sequences by test transcription factors. Screening of transcription factor-binding sequences was performed by the following steps (Ishihama 2010): preparation of DNA fragments of the *E. coli* genome DNA; affinity isolation of DNA-His-tagged transcription factor complexes; sequence determination of transcription factor-associated DNA fragments by either cloning-sequencing (SELEX-clos) or mapping using DNA tilling array (DNA-chip).

One of the marked findings in this series of Genomic SELEX screening is that the number of regulation targets by a single transcription factor is more than those so far identified. For instance, the number of regulation targets increased 2.3 fold (345/150) for CRP (cAMP receptor protein) (Shimada et al. 2011a), and 8.9 fold (169/19) for Cra (catabolite repressor and activator) (Shimada et al. 2011b). After detailed analysis of the regulation target genes, it became clear that these two transcription factors, CRP and Cra, are the key global regulators of the whole set of genes for transport and utilization of carbon sources.

1.3 Mater Regulator for Single-cell Planktonic Growth

Motility and chemotaxis of bacteria allow cells to move away from stressful environments. Motility of *E. coli* in batch cultures, as measured by cell swimming speed, which is low in early-exponential-phase cells, peaks in the late exponential phase, and declines upon transition into the early stationary phase (Amsler et al. 1993). The increase in swimming speed may be correlated with the increase in the number of flagella per cell (and the flagellar length). The decrease in speed may also be correlated with the decrease in torque produced by the flagellar motors, reflecting decrease in the proton motive force. The expression, synthesis, assembly and function of flagella and motility requires the expression of more than 50 genes, constituting a large and co-ordinately regulated flagella regulon (Macnab 1996), which can be classified into three groups altogether forming the regulation cascade (Kutsukake et al. 1990, Claret and Hughes 2002).

The genes for flagella formation are considered to be under a complex regulation system. FlhDC complex is the master regulator controlling the hierarchy of regulation network for flagella synthesis (Claret and Hughes 2002, Stafford et al. 2005). RpoF, one of the seven *E. coli* sigma factors, is synthesized under the control of FlhDC, and participates in transcription of the genes downstream of the flagella regulon. The increase in flagella number in the late exponential phase is, at least in part, due to the increase in RpoF level, and the decrease in flagella number in the stationary phase is correlated with the decrease in the level of RpoF (Jishage et al. 1996). The decrease in RpoF takes place in parallel with the increase in RpoS, which is needed for transcription of some stationary phase-specific genes (Ishihama 1999, 2000). Among the seven sigma factors in *E. coli*, two minor sigma factors, RpoN and RpoF, are present, besides the major RpoD sigma, in exponential phase cells (Jishage et al. 1996). Upon entry to the stationary phase, unused RpoD sigma is converted into a stored inactive form by binding the anti-sigma factor Rsd (regulator of sigma D) (Jishage and Ishihama 1998). Likewise RpoF is stored as an inactive form as a complex with FlgM, indicating that the level of functional RpoF sigma is controlled by both the synthesis and the activity of RpoF protein. Activation of the pre-existing RpoF by dissociation of its anti-sigma factor FlgM may also be involved in the increase in flagella production. The expression of the *flhDC* master operon is under the control of multiple transcription factors, each sensing an environmental signal or condition, including CsgD, the master regulator of biofilm formation.

After Genomic SELEX screening of CsgD, we identified multiple CsgD-binding sites within the intragenic spacer between divergently transcribed *fliE* and *fliFGHIJK* operons (Ogasawara et al. 2011). The *fliE* gene encodes a component of the flagella basal body assembly connecting MS ring and rod, while the *fliFGHIJK* operon encodes essential components and functional modulators of flagella, including components of flagella export apparatus (Minamino et al. 2009, Muller et al. 1992). Since both the *fliE* and *fliF* promoter activities markedly increase in the *csgD* mutant, we proposed that CsgD is involved as a repressor of the flagella regulon. Interestingly this CsgD-binding site overlaps with the binding site of FlhDC, the master regulator of genes for flagella formation. Thus the apparent repression of *fliF* promoter by CsgD may be attributable to interference of FlhDC binding to its target site (for details see below).

1.4 Master Regulator for Biofilm Formation

Under stressful conditions, the growth habit of bacteria changes from single-cell planktonic growth to assembled biofilm. Formation of a biofilm begins with the attachment of free-floating microorganisms on to a solid surface and cell-cell contact using adhesion structures such as curli fimbriae. Expression of the *csgBAC* and *csgDEFG* operons for the synthesis, secretion and assembly of curli bimbriae is under the positive control of CsgD (Curlin subunit gene D), a FixJ/LuxR/UhpA-family transcription factor (Brombacher et al. 2003, Hammar et al. 1995, Prigent-Combaret et al. 2001). Changes in various environmental conditions such as low osmolarity, low temperature, starvation and high cell density, influence directly or indirectly, the expression of CsgD (Boughdour et al. 2004, Brombacher et al. 2003). Reflecting the responsibility of CsgD expression to various external stresses, a number of transcription factors, each monitoring a different factor or condition, are involved in the regulation of *csgD* promoter (Ishihama 2010, Ogasawara et al. 2010a, 2010b).

In order to get an insight to the whole network of gene regulation involving CsgD, the identification of whole set of regulation targets by CsgD is an initial and essential step. In the absence of knowledge of effector affecting CsgD activity, we searched the regulation targets by analysis of CsgD-associated genes *in vivo* using the ChIP-chip (chromatin immunoprecipitation and microarray) system (Grainger and Busby 2008,

Grainger et al. 2005, Shimada et al. 2008). As a result, we have identified more than 20 CsgD-binding sites along the *E. coli* genome (Ogasawara et al. 2011). CsgD was found to modulate the expression of not only the *csgBAC* and *csgDEFG* operons encoding the components and assembly of curli, and its regulator CsgD (Hammar et al. 1995, Ogasawara et al. 2011), but also a set of genes for adaptation of cell physiology to the biofilm life style (Brombacher et al. 2006, Chirwa and Herrington, 2003, Ogasawara et al. 2011), including *adrA* encoding diguanylate cyclase for synthesis of cyclic di-GMP (Ogasawara et al. 2011), a bacterial second messenger (Simm et al. 2004) for enhancement of cellulose production (Chirwa and Herrington 2003, Pesavento et al. 2008, Zogaj et al. 2001). c-di-GMP induces the biofilm formation and inhibits flagella production and rotation (Armitage and Berry 2010, Boehm et al. 2010, Wolfe and Visick 2008). Using the promoter assay using *lacZ* reporter, we found that a group of promoters were activated in the *csgD* mutant, indicating that CsgD is involved as a repressor for this group of promoters. The activity of promoters carrying the CsgD-binding site at upstream of the respective promoters decreased in the mutant lacking CsgD, indicating the involvement of activator CsgD. Between the divergently transcribed *csgDEFG* and *csgBC* operons, a total of seven CsgD-binding sites were identified and in addition, a number of different transcription factors were found to bind to the same regions (Ogasawara et al. 2010a, 2010b). Thus the expression of CsgD should be under a complex regulation system involving the molecular interplay between different CsgD molecules and between CsgD and other transcription factors (Fig. 9.3).

1.5 Cross-regulation between Flagella Synthesis and Biofilm Formation

The biofilm master regulator CsgD binds to the intragenic spacer region between the *csgDEFG* and *csgBC* operons and activates the genes for the formation of curli fimbriae, the essential cell structure for cell adhesion on solid surface and cell-cell association, but here we found that the master regulator of flagellar formation, FlhDC, binds to the same region and represses the *csg* operons, thereby inhibiting biofilm formation (Fig. 9.3). On the other hand, CsgD binds to the intergenic spacer between the *fliE* and *fliFGHIJK* operons, where the master regulator FlhDC of flagellar formation binds and activates the genes related to flagellar formation and rotation, and represses the *fli* operons (Fig. 9.3). Interestingly the binding site of CsgD in the intergenic spacer between the *fliE* and *fliFGHIJK* operons overlaps with the binding site of FlhDC complex, the master regulator of flagella formation. Likewise the binding site of FlhDC within the intragenic spacer between the *csgDEFG* and *csgBC* operons overlaps with the binding sites of CsgD in this region. Thus we propose the competition

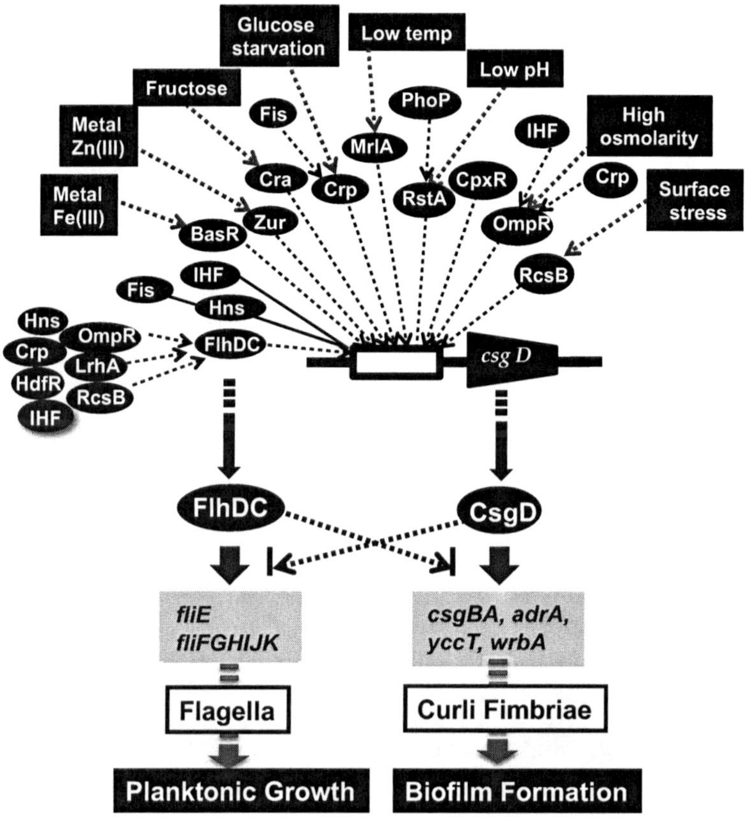

Figure 9.3: Cross-talk of the regulation network between flagella synthesis and biofilm formation. The promoter for *csgD* encoding the master regulator of biofilm formation is under the control of more than 10 transcription factors (Ogasawara et al. 2010a, 2010b), each sensing a different environmental factor of condition, including the master regulator FlhDC for the flagella regulon. One of the regulation targets of CsgD is the flagella regulon that plays a key role for single-cell planktonic growth (Ogasawara et al. 2011).

between two key regulators, CsgD and FlhDC, that controls the switching between two bacterial habits, single-cell planktonic growth and assembled biofilm (Fig. 9.3). In agreement with this competition model of two master regulators, the binding consensus sequence is similar between these two transcription factors.

In addition to the *fliE-fliFGHIJK* intergenic spacer region, CsgD associates at the intergenic regions between *flgB* and *flgA* within the *flgBAMN* operon and between *fliC* and *fliD* within the *fliCD* operon, and on the *fliL* gene within the *fliFGHIJK* operon, altogether leading to inhibition of flagella formation and assembly in good concert with the proposal that CsgD is involved in regulation of flagella-related operons (Zakikhany et al. 2010, Ogasawara et al. 2011). Along this line, c-di-GMP, the product of CsgD-activated AdrA diguanylate cyclase, is an inhibitor of cell motility (Pesavento et al. 2008), by interfering flagellar motor speed through c-di-GMP-binding protein YcgR. Accordingly CsgD interferes with the formation and function of flagellar, leading to inhibition of planktonic growth for switching toward biofilm formation.

2. Detection and Fractionation of Heterogeneity in Cell Populations under Laboratory Culture Conditions

2.1 Growth Phase-coupled Differentiation of E. coli Cells

When *E. coli* cells switch the life mode from single planktonic cell growth to biofilm mode, the flagella formation is turned off and in turn the production of curli fimbriae and extracellular polysaccharides for attachment onto solid surface and cell-cell adhesion is switched on (Barnhart and Chapman 2006, Chapman et al. 2002). Sometime ago we realized that even in laboratory culture conditions, *E. coli* cells are heterogenous in population, including various types of cells, each differing in morphology and physiology (Makinoshima et al. 2002, 2003). After extensive trials, we succeeded in fractionating random *E. coli* cultures into apparently homogenous cell populations by using Percoll gradient centrifugation (Makinoshima et al. 2002). Transition from exponential growth phase to stationary phase is not synchronized in random cultures, thus generating heterogeneous populations containing various types of cells, each differing in cell shape, physical properties such as buoyant density, and molecular composition (Makinoshima et al. 2002, 2003). Such variation in cell populations may arise not only from the fluctuation in the cell division during transition from the resting to growing phase and/or from the exponential growth to stationary phase (Makinoshima et al. 2002, 2003) but also from the variation of time duration in the stationary phase prior to the restart of cell growth (Pin and Baranyi 2008). Up to the present moment, however, most parameters of bacterial cultures, including cell growth, cell division, genome expression, promoter activity and RNA and protein levels, have been measured as the average of cell populations in random cultures, relying on the belief that bacteria are homogeneous in population. Thus, the findings of population heterogeneity have raised criticism over the established concept of the regulation of *E. coli* promoters that have been determined as the average of a number of cells, each exhibiting a different pattern of the genome expression.

2.2 Percoll Density Gradient Centrifugation for Separation of Differentiated Cells

Percoll is colloidal silica that is coated with polyvinylpyrrolidone to protect the cells from toxic effects. Since the viscosity is lower than other reagents to give the same densities, materials as large as marker beads and bacterial cells quickly sediment to positions characteristic of their densities. By Percoll centrifugation, random cultures of *E. coli* were fractionated into more than 20 populations, each having a different bouyant density and apparently representing a specific stage of cell differentiation from exponential growth to stationary phase (Makinoshima et al. 2002) (Fig. 9.4).

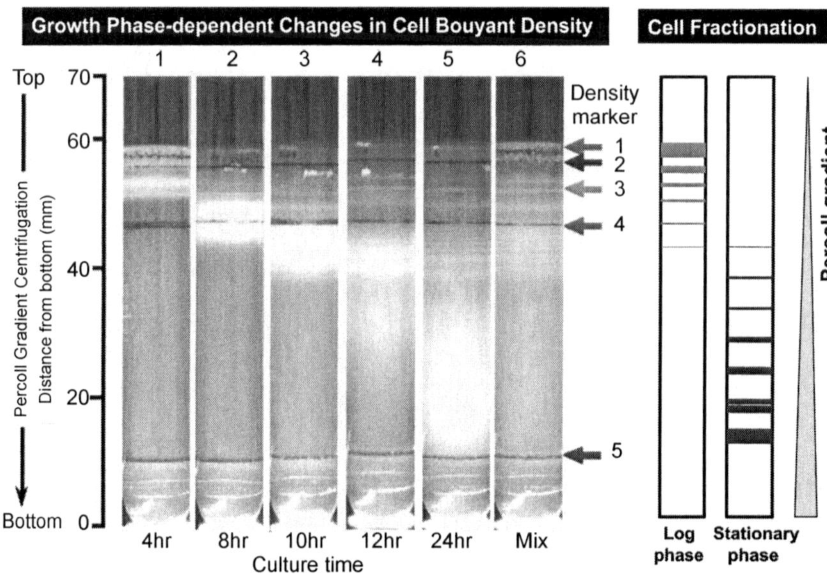

Figure 9.4: Bacterial cell fractionation through Percoll gradient centrifugation. Random culture of *E. coli* can be fractionated by Percoll density gradient centrifugation (Makinoshima et al. 2002). Upon entry into the stationary growth phase, the cell buoyant density increases step-wise, thereby allowing the isolation of homogeneous cell population. In each step of the density increase, a different gene(s) is involved.

Color image of this figure appears in the color plate section at the end of the book.

The bacterial biomass must be more variable than previously assumed from volume-based measurements (Loferer-Krossbacher et al. 1998). The density increase was found to be impaired at an early step for a mutant

E. coli with the disrupted *rpoS* gene, which encodes the RNA polymerase RpoS (sigma-S) for stationary-phase gene transcription. This finding suggests that an as yet unidentified gene(s) under the control of RpoS is needed for the initial stage of cell density increase. The major factor for the success of cell separation by the newly developed method is the increase in cell buoyant density or weight-to-volume ratio during transition from the exponential growing to stationary cells. Transmission electron microscopy and densitometric image analysis of cell thin sections indicated the accumulation of electron-dense areas (Makinoshima et al. 2003), which takes place in parallel with the change in the chemical composition of cells such as the increase in membrane cyclopropyl derivatives by replacing unsaturated fatty acids, the increase in compatible solutes with osmo-protection activity such as trehalose and glycine betaine, the accumulation of storage compounds such as glycogen and polyphosphate, the decrease in polyamines, and the decrease in free water content (Dubochet et al. 1983, Roszak and Colwell 1987, Kolter et al. 1993, Ishihama 1999). Polyphosphate, the storage material for excess ATP (Kornberg et al. 1999) plays a regulatory role in the function of RNA polymerase sigma S (Kusano and Ishihama 1997) and an as yet unidentified function such as detoxification by sequestering heavy metals forming the electron-dense granule of metal-sequestered polyphosphate (Pan-Hou et al. 2002). Polyhydroxyalkanoic acids (PHA) are carbon and energy storage polymers that accumulate in inclusion bodies in many bacteria in response to environmental conditions. The PHA inclusion bodies are not only a source of carbon and energy but also plays an anchoring platform for hydrophilic proteins, such as PhaP, a bacterial storage protein (McCool and Cannon 1999). In addition, the spatial rearrangement of cell components within the stationary-phase *E. coli* cells such as the compaction of nucleoid by replacing associated proteins from log-phage specific Fis to stationary-phase specific Dps (Azam et al. 1999), and the decrease in associated polyamines (Yoshida et al. 2002) might result in cell density change.

3. Establishment of Single Cell Culture Systems and Determination of Promoter Activity within Single Cells

3.1 Construction of Cell Chips

We have made a systematic effort in order to make a breakthrough for solving the cell heterogeneity problem that arises from the population heterogeneity in random bacterial cultures. Our initial effort focused on fractionation of bacterial cells by density gradient centrifugation (Makinoshima et al. 2002, 2003) or by FACS based on the intensity of expression levels of reporter fluorescent proteins (Roostalu et al. 2008, Teramoto et al. 2010b). Our recent effort, however, has shifted to the development of experimental systems to analyze the genome regulation within a single cell. This report describes our initial attempts of cell chip construction and measurement of the promoter strength within single *E. coli* cells.

In this study, we constructed and used two types of the cell chip for immobilization of portions of bacterial culture and control of the microenvironment of immobilized cells by circulating the culture medium through micro-channels (Fig. 9.5). The transparent silicone elastomer, poly(dimethylsiloxane) (PDMS), was used for construction of the cell chip with two micro-channels. A test bacterial culture containing appropriate concentrations of poly-(*N*-isopropylacrylamide) (PNIPAAm) was prepared at 25°C (the chamber temperature) and passed through the cell chips at a constant flow rate from the input channel (inlet A). For conversion of a portion of the bacterial culture into gel, the transparent microelectrodes of ITO (indium tin oxide) were inserted into the cell chip (Arai et al. 2003, Yamanishi et al. 2009, Teramoto et al. 2010a). In type-1 chip, three ITO electrodes were inserted for quick equibration of the local temperature while in type-2 chip, a single electrode was used for warming up the local temperature. Upon warming up a local region within the cell to 30°C, a portion of bacterial culture was converted to gel for immobilization of a group of cells (Ichikawa et al. 2005). For prolonged culture of immobilized bacterial cells within the cell chip under the same conditions, a fresh medium was continuously circulated through inlet channel B. As a result, we succeeded in following the growth and fate of immobilized cells for a prolonged period. To test possible influence of effectors on the promoter activity, appropriate inducers or co-repressors can be added into the circulating medium.

Formations of PNIPAAm gels are concentration- and temperature-dependent. For long-term culture of immobilized bacterial cells within the gels, it is critical to maintain constant temperature. For this purpose, the temperature sensor was attached in an improved version of the cell chip so as to measure and control the gel size and width. Acrylamides at high concentrations are toxic for bacterial growth. The growth rate decreased concomitant with the increase in the concentration of poly-(*N*-isopropylacrylamide) [PNIPAAm]. Below 10% PNIPAAm, cell growth was virtually unaffected, and thus PNIPAAm was added at the final concentration of 10% for effective gelation of *E. coli* culture after raising the temperature above 30°C. PNIPAAm gels are thermo-reversible, and could be resolubilized by decreasing the cell-chip temperature to 20°C. The survival of *E. coli* cells immobilized within the gel could be measured after resolubilization. More than 90–95% *E. coli* cells were viably recovered after immobilization for 1 hr within the 10% PNIPAAm gel.

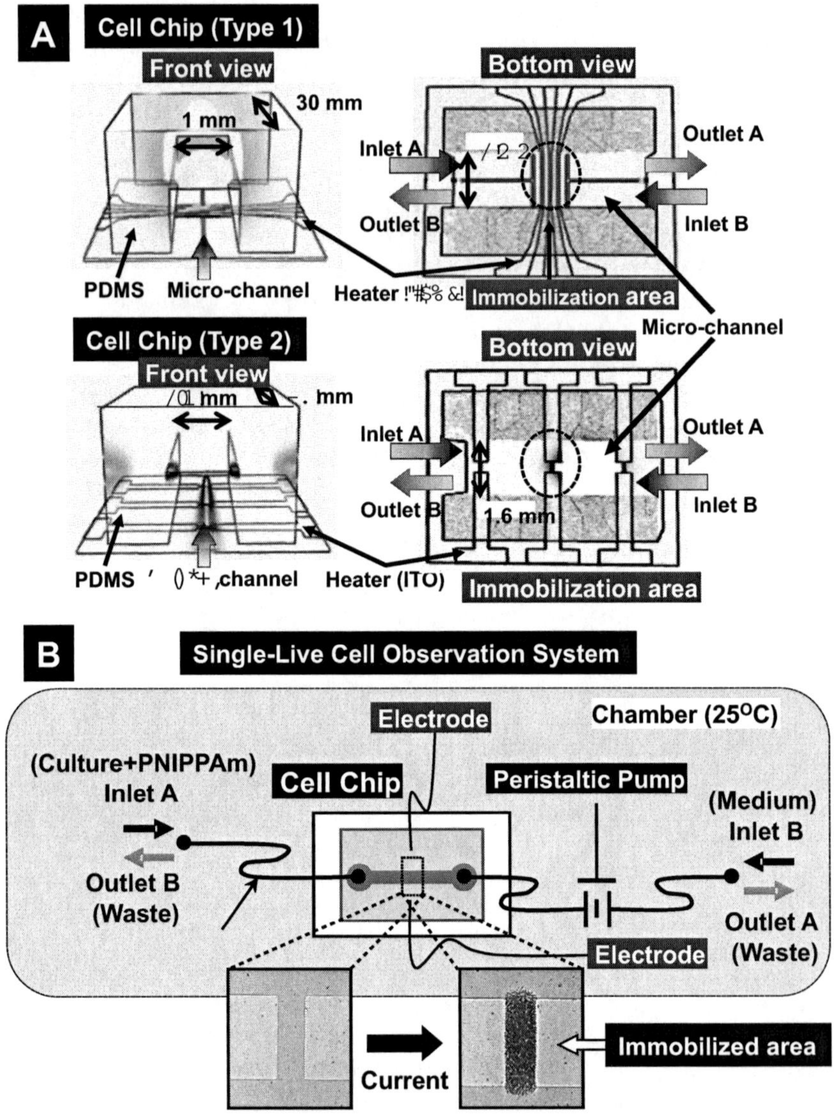

Figure 9.5: Observation system of single live bacterial cells. For monitoring gene expression in individual *E. coli* cells, a portion of cell culture was immobilized within a hand-made cell chip. The fate of individual cells can be followed for prolonged time periods. Using the singe-cell observation system, the strength and regulation of test promoters were monitored in the presence and absence of inducers (Teramoto et al. 2010a).

3.2 Real-time Single-cell Assay of Promoter Strength and Regulation

The newly developed cell-chip devices were first applied for quantitative determination of the activity and regulation of *E. coli* promoters within single live-bacterial cells. For this purpose, we used the promoter assay system using the two-fluorescent reporter vector pGRP (Shimada et al. 2004). DNA segments containing the test promoters have been PCR-amplified and inserted into pGRP at the site immediately upstream of GFP (green fluorescent protein) coding sequence. Each promoter segment covered from the translation initiation codon to approximately 300 bp upstream of the transcription initiation sites or to about 500 bp upstream of the initiation codon (in the case the transcription initiation site has not been identified). From the same vector, RFP (red fluorescent protein) can be expressed under the control of reference promoter *lacUV5*. The activity *in vivo* of test promoters can be easily but accurately determined simply by measuring the GFP/RFP ratio, thereby avoiding the fluctuation of reporter gene expression arising from changes in plasmid copy number.

The promoter activity determined with *E. coli* liquid cultures represents the average value of a number of cells. Attempts were then made to determine the time-dependent change of the promoter activity in individual cells within the same culture. After immobilization of a portion of the *E. coli* culture within the cell chip, the cell culture was continued by constant circulation of fresh LB medium around the PNIPAAm gel. The promoter activity can be determined by measuring GFP expression level in immobilized individual cells. For instance, using the single-cell observation system, the promoter of *gcl* gene, that encodes glyoxylate carboligase in the pathway of purine degradation (Hasegawa et al. 2008), was detected as early as 2 min after the addition of an inducer, glyoxylate, and increased thereafter almost linearly up to 30 min. The increase of GFP level could be followed at least up to 10 hr.

3.3 Fluctuation of the Promoter Strength between Cells within Random Liquid Culture

One unexpected finding obtained using the single-cell observation system is that: the difference in GFP level among cell populations was sometimes more than 10 fold; the pattern of time-dependent increase in GFP level is also different between individual cells; and the expression of GFP was not detected for approximately 30–40%. One possibility of this unexpected observation is that a large number of immobilized cells are not viable within the gel, but this possibility was excluded because most of the immobilized cells continued to grow after resolubilizing the gel by decreasing the temperature. We then predicted that the *gcl* promoter is not expressed in a considerable population of the *E. coli* cells used.

In parallel, we also examined possible fluctuation of the *gcl* promoter activity in random LB liquid culture by using Flow cytometry, which enables the identification of the distribution of GFP fluorescence among a large number of cells within a single liquid culture. Fluorescence of GFP in individual bacterial cells was recorded simultaneously using laser excitation and light detection. The peak position of fluorescent cells increased concomitantly with the increase of culture time after the addition of glyoxylate. The width of the distribution of fluorescent cells also increased concomitant with the increase of incubation time after glyoxylate addition, indicating the time-dependent increase in the fluctuation level in GFP expression. The number of fluorescent cells increased up to approximately 50 and 65% at 30 and 60 min, respectively, after the addition of glyoxylate. Even after 60 min incubation in the presence of inducer gloxylate, about one-third of the cells did not express detectable level of GFP, in good agreement with the single-cell assay. Most of these non-fluorescent cells were, however, stained with DAPI (4',6-diamino-2-phenylindole) but not with PI (propidium iodide), indicating that the cells without expressed GFP are alive.

3.4 Single-cell Observation System of Prokaryotes

At the initial stage of single bacterial cell biology, attempts were made to characterize individual bacterial cells within colonies. For instance, the quantitative relation between transcription factor concentrations and the level of protein production within a single *E. coli* cell was analyzed by measuring the level of lambda repressor and a protein product under the control of the repressor using fluorescent time-lapse microscopy of individual cells within micro-colonies (Rosenfeld et al. 2005). The cells within colonies are, however, not isolated, being in close contact with neighboring cells. Attempts were then made to isolate GFP-expressing *E. coli* cells from RFP-expressing *Pseudomonas putida* using an inverted microscope, a microinjector and a micromanipulator (Ishoy et al. 2006). Isolated *E. coli* cells were captured in a microcapillary for further characterization. Isolation of single cells by micromanipulation is not useful for characterization of the response of a large number of isolated cells after changes in its environmental condition.

The two-step method, initial immobilization of *E. coli* cells on a glass coverslip coated with anti flagellin antibodies followed by covering the fixed cells with low-melting agarose, was employed for characterizing the short-time dynamics of specific RNA levels within a single cell (Le et al. 2005). Using a flow chamber, *E. coli* cells were attached to a transparent solid surface and the growth of fixed individual cells was monitored by removing the divided daughter cells by shear force of the flow (Elfwing et al. 2004, Pin and Baranyi 2006). Fluctuation in the time lag for entry of stationary-phase *E. coli* cells into growing phase was monitored using this system (Pin and Barany 2008). On-chip micro-culture system with a cell chamber separated from circulating medium channel was developed for observation of the response of isolated individual cells to changes in environment (Inoue et al. 2001). Using this system, the origin of individuality of two daughter cells born from an isolated single *E. coli* mother cell was examined by monitoring the change of swimming behavior and tumbling frequency (Umehara et al. 2007). Using a microfuidic bioreactor that enabled long-term culture, small populations of bacteria with singe-cell resolution were monitored for analysis of the mechanism of population density control (Balagadde et al. 2005). Single-cell protein production dynamics have been investigated by measuring the induction of fluorescent proteins within microfluidic devices (Gefen et al. 2008). Taking advantage of lithography, a micro-Petri dish has been developed for a high-through put screening (Ingram et al. 2007), but this miniaturized culture system can be used for the purpose of single cell observation. However, with the use of experimental systems for monitoring single cells in liquid cultures, it is not possible to follow the fate of single specific cells for a prolonged time period. We then switched our effort to develop the fixed cell system by using the reversible immobilization gel.

Culture of cells immobilized on the surface of beads was used to screen enzyme mutants with altered activities by microscopic observation using fluorescent substrates (Freeman et al. 2003). Single *E. coli* cells were detected in a Y-channel polydimethylsiloxane (PDMS) microfluidic device through optical fiber monitoring of latex immunoagglutination with use of polystyrene particles conjugated with polyclonal anti-*E. coli* (Han et al. 2008). For the time-lapse fluorescence microscopic analysis of the induction kinetics of a specific promoter, *E. coli* cells with reporter GFP under the control of *ara* promoter were introduced into a microfluidic change with poly-L-lysine-coated chamber and the GFP expression was monitored for individual cells after addition of arabinose (Megerle et al. 2008). We then developed another system of single live-bacterial cell analysis by employing a thermo-reversible hydrogel without using surface coating techniques. Since aliquots of the culture were immobilized within the gel without interfering with the medium flow outside the gel, this system allows the change in environmental conditions such as the addition of inducer or co-repressor and the changes in nutrients, salts, pH, osmolarity and temperature. Preliminary attempts have been shown to be successful for monitoring the promoter activity of single target cells for prolonged time periods. Application of this newly developed experimental system for analysis of uncharacterized *E. coli* promoters is in progress.

In conclusion, we have developed a new real-time single-bacterial cell assay system of the promoter strength and regulation dynamics and successfully employed for monitoring the *E. coli gcl* promoter activity in single target cells in the presence and absence of effectors. Surprisingly in the case of *gcl p*romoter, the fluctuation of the promoter activity between cells within a single and the same culture was remarkable. The wide-range variation in the *gcl* promoter activity is not due to the effect of the gel, because this heterogeneity in cell populations with respect to the *gcl* promoter expression was confirmed after FACS analysis of the liquid culture. This finding supports our prediction that the cell heterogeneity exists even in laboratory bacterial cultures. The newly developed single-cell observation system could be useful for detailed characterization of individual bacterial cells.

CONCLUSION

Under laboratory culture conditions, bacteria grow as apparently homogenous single-cell planktonic states, but in nature, bacteria survive in the assembled biofilm, in which each cell plays a unique role different from other surrounding cells. For switching control in bacterial habitat between single-cell planktonic growth and assembled biofilm, the competition between two key transcription factors, the master regulator CsgD of biofilm formation and the master regulator FlhDC of flagella formation in binding to the common and shared promoters was found to play a crucial role. To solve the problem that arises from cell heterogeneity, we developed two stages of the experimental system: the isolation of homogenous cell populations through Percoll gradient centrifugation; and the development of single-cell observation systems using newly constructed cell-chips. The activity and regulation of promoters within single *E. coli* cells were observed in the presence or absence of effectors for prolonged time periods.

REFERENCES

Amsler, C.D. and M. Chao and P. Matsumura. 1993. Multiple factors underlying the maximum motility of *Escherichia coli* as cultures enter post-exponential growth. J Bacteriol 175: 6238–6244.

Arai, F. and A. Ichikawa, T. Fukuda and T. Kutsuragi. 2003. Isolation and extraction of target microbes using thermal sol-gel transformation. Analyst 128: 547–551.

Armitage, J.P. and R.M. Berry. 2010. Time for bacteria to slow down. Cell 141: 24–26.

Azam, T.Z. and A. Iwata, A. Ueda and A. Ishihama. 1999. Growth phase-dependent variation in the protein composition of *Escherichia coli* nucleoid. J Bacteriol 181: 6361–6370.

Balagadde, F.K. and L.You, C.L. Hansen, F.H. Arnold and S.R. Quake. 2005. Long-term monitoring of bacteria undergoing programmed population control in a microchemostat. Science 309: 137–140.

Barnhart, M.M. and M.R. Chapman. 2006. Curli biogenesis and function. Annu Rev Microbiol 60: 131–147.

Beloin, C. and J. Valle, P. Latour-Lambert, P. Faure, J. Kzreminski, D. Balestriono, J.A. Haagensen, S. Molin, G. Prensier, B. Arbeille and J.M. Ghigo. 2004. Global impact of mature biofilm lifestyle on *Escherichia coli* K-12 gene expression. Mol Microbiol 51: 659–674.

Boehm, A. and M. Kaiser, H. Li, C. Spangler, C.A. Kasper, C.M. Ackermann, V. Kaever, V. Sourjik, V. Roth and U. Jenal. 2010. Second messenger-mediated adjustment of bacterial swimming velocity. Cell 141: 107–116.

Bourgdour, A. and C. Lelong and J. Geiselmann. 2004. Crl, a low temperature-induced protein in *Escherichia coli* that binds directly to the stationary phase sigma subunit of RNA polymerase. J Biol Chem 279: 19540–19550.

Brombacher, E. and C. Dorel, A.J. Zehnder and P. Landini. 2003. The curli biosynthesis regulator CsgD co-ordinates the expression of both positive and negative determinants for biofilm formation in *Escherichia coli*. Microbiology 149: 2847–2857.

Brombacher, E. and A. Baratto, C. Dorel and P. Landini. 2006. Gene expression regulation by the curli activator CsgD protein: Modulation of cellulose biosynthesis and control of negative determinants for microbial adhesion. J Bacteriol 188: 2027–2037.

Claret, L. and C. Hughes. 2002. Interaction of the atypical prokaryotic transcription activator FlhD2C2 with early promoters of the flagellar gene hiererchy. J Mol Biol 321: 185–199.

Casadaban, M.J. 1976. Transcription and fusion of the *lac* genes to selected promoters in *Escherichia coli* using bacteriophage lambda and Mu. J Mol Biol 104: 541–555.

Chapman, M.R. and L.S. Robinson, J.S. Pinkner, R. Roth, J. Heuser, M. Hammar, S. Normark and S.J. Hultgren. 2002. Role of *Escherichia coli* curli operons in directing amyloid fiber formation. Science 295: 851–855.

Chirwa, N. and M.B. Herrington. 2003. CsgD, a regulator of curli and cellulose synthesis, also regulates serine hydroxymethyltransferase synthesis in *Escherichia coli* K-12. Microbiology 149: 525–535.

Claret, L. and C. Hughes. 2002. Interaction of the atypical prokaryotic transcription activator FlhD2C2 with early promoters of the flagellar gene hierarchy. J Mol Biol 321: 185–199.

Dubochet, J. and A.W. McDowall, B. Menge, E.N. Schmid and K.G. Lickfeld. 1983. Electron microscopy of frozen hydrated bacteria. J Bacteriol 155: 381–390.

Elfwing, A. and Y. LeMarc, J. Baranyi and A. Ballagi. 2004. Observing growth and division of large numbers of individual bacteria by image analysis. Appl Environ Microbiol 70: 676–678.

Freeman, A. and N. Cohen-Hadar, S. Abramov, R. Modai-Hod, Y. Dror and G. Georgiou. 2003. Screening of large protein libraries by the "cell immobilized on adsorbed bead" approach Biotechnol Bioengineer 86: 196–200.

Gefen, O. and C. Gabay, M. Mumcuoglu, G. Engel and N.Q. Balaban. 2008. Single-cell protein induction dynamics reveals a period of vulnerability to antibiotics in persister bacteria. Proc Natl Acad Sci USA 105: 6145–6149.

Gerstel, U. and C. Park and U. Romling. 2003. Complex regulation of *csgD* promoter activity by global regulatory proteins. Mol Microbiol 49: 639–654.

Grainger, D.C. and S.J. Busby. 2008. Methods for studying global patterns of DNA binding by bacterial transcription factors and RNA polymerase. Biochem Soc Trans 36: 754–757.

Grainger, D.C. and D. Hurd, M. Harrison, J. Holdstock and S.J. Busby. 2005. Study of the distribution of *Escherichia coli* cAMP-receptor protein and RNA polymerase along the *E. coli* chromosome. Proc Natl Acad Sci USA 102: 17693–17698.

Hammar, M. and A. Arnqvist, Z. Bian, A. Olsen and S. Normark. 1995. Expression of two *csg* operons is required for production of fibronectin- and congo red-binding curli polymers in *Escherichia coli* K-12. Mol Microbiol 18: 661–670.

Han, J-H. and B.C. Heinze and J-Y. Yoon. 2008. Single cell level detection of Escherichia coli in microfluidic device. Biosens Bioelectron 23: 1303–1306.

Hasegawa, A. and H. Ogasawara, A. Kori, J. Teramoto and A. Ishihama. 2008. AllR is the allantoin/glyoxylate-sensing master regulator of the genes for degradation and reutilization of purines. Microbiology 154: 3366–3378.

Ichikawa, A. and F. Arai, K. Yoshikawa, T. Uchida and T. Fukuda. 2005. *In situ* formation of a gel microbead for indirect laser micromanupulation of microorganisms. Appl Phys Lett 87: 191108.

Ingham, C.J. and A.D. Sprenkels, J. Bomer, D. Molenaar, A. van den Berg, J.E.T. van Hylckama Vlieg and W.M. de Vos. 2007. The micro-Petri dish, a million-well growth chip for the culture and high-throughput screening of microorganisms. Proc Nat Acad Sci USA 104: 18217–18222.

Inoue, I. and Y. Wakamoto, H. Moriguchi, K. Okano and K. Yasuda. 2001. On-chip culture system for observation of isolated individual cells. Lap on a Chip 1: 50–55.

Ishihama, A. 1990. Molecular assembly and functional modulation of *Escherichia coli* RNA polymerase. Adv Biophys 26: 19–31.

Ishihama, A. 1999. Modulation of the nucleoid, the transcription apparatus, and the translation machinery in bacteria for stationary phase survival. Genes Cells 3: 135–143.

Ishihama, A. 2000. Functional modulation of *Escherichia coli* RNA polymerase. Annu Rev Microbiol 54: 499–518.

Ishihama, A. 2009. The nucleoids. *In:* A. Boek, R. Curtiss III, JB Kaper, PD. Karp FC Neidhardt, T. Nystrom, J.M. Slauch, C.L. Squires and D Ussery [eds]. EcoSal—Escherichia coli and Salmonell: Cellular and Molecular Biology. ASM Press, Washington

Ishihama, A. 2010. Prokaryotic genome regulation: Multi-factor promoters, multi-target regulators and hierarchic networks. FEMS Microbiol Rev 34: 628–645.

Ishoy, T. and T. Kvist, T.P. Westermann and B.K. Ahring. 2006. An improved method for single cell isolation of prokaryotes from meso-, thermo- and hyperthemophilic environments using micromanipulation. Appl Micriobiol Biotechnol 69: 510–514.

Jishage, M. and A. Ishihama. 1995. Regulation of RNA polymerase sigma subunit synthesis in *Escherichia coli*: Intracellular levels of σ70 and σ38. J Bacteriol 177: 6832–6835.

Jishage, M. and A. Ishihama. 1998. A stationary phase protein in *Escherichia coli* with binding activity to the major σ subunit of RNA polymerase. Proc Natl Acad Sci USA 95: 4953–4958.

Jishage, M. and A. Iwata, S. Ueda and A. Ishihama. 1996. Regulation of RNA polymerase sigma subunit synthesis in *Escherichia* coli: Intracellular levels of four species of sigma subunit under various growth conditions. J Bacteriol 178: 5447–5451.

Kolter, R. and D.A. Siegele and A. Tormo. 1993. The stationary phase of the bacterial life cycle. Ann Rev Microbiol 47: 855–874.

Kornberg, A. and N.N. Rao and D. Ault-Riche. 1999. Inorganic polyphosphate: a molecule of many functions. Annu Rev Biochem 68: 89–125.

Kusano, S. and A. Ishihama. 1997. Functional interaction of *Escherichia coli* RNA polymerase with inorganic polyphosphate. Genes Cells 2: 433–441.

Kutsukake, K. and Y. Ohya and T. Lino. 1990. Transcriptional analysis of the flagellar regulon of Salmonella typhimurium. J Bacteriol 172: 741–747.

Le, T.T. and S. Harlepp, C.C. Guet, K. Dittmar, T. Emonet, T. Pan and P. Cluzel. 2005. Real-time RNA profiling within a single bacterium. Proc Natl Acad Sci USA 102: 9160–9164.

Liu, X. and P. Matsudaira. 1994. The FlhD/FlhC complex, a transcriptional activator of the Escherichia coli flagella class II operons. J Bacteriol 176: 7345–7351.

Loferer-Krossbacher, M. and J. Klima and R. Psenner. 1998. Determination of bacterial cell dry mass by transmission electron microscopy and densitometric image analysis. Appl Environ Microbiol 64: 688–694.

Macnab, R.M. 1996. Flagella and motility. pp. 123–145. *In:* F.C. Neidhardt et al. [eds.]. Escherichia coli and Salmonella. American Society for Microbiology Press, Washington, D.C.

Makinoshima, H. and A. Nishimura and A. Ishihama. 2002. Fractionation of *Escherichia coli* cell populations at different stages during growth transition to stationary phase. Mol Microbiol 43: 269–279.

Makinoshima, H. and S. Aizawa, H. Hayashi, T. Miki, A. Nishimura and A. Ishihama. 2003. Growth phase-coupled alterations in cell structure and function of *Escherichia coli*. J Bacteriol 185: 1338–1345.

McCool, G.J. and M.C. Cannon. 1999. Polyhydroxyalkanoate inclusion body-associated proteins and coding region in *Bacillus megaterium*. J Bacteriol 181: 585–592.

Megerle, J.A. and F. Georg, U. Gerland, K. Jung and J.O. Raeder. 2008. Timing and dynamics of single cell gene expression in the arabinose utilizing system. Biophys J 95: 2103–3115.

Miller, J.H. 1972. Experiments in Molecular Genetics. Cold Spring Harbor Laboratory Press, New York.

Minamino, T. and N. Moriya, T. Hirano, K.T. Hughes and K. Namba. 2009. Interaction of *fliK* with the bacterial flagellar hook is required for efficient export specificity switching. Mol Microbiol 74: 239–251.

Muller, V. and C.J. Jones, I. Kawagishi, S. Aizawa and R.M. Macnab. 1992. Characterization of the *fliE* genes of *Escherichia coli* and *Salmonella typhimurium* and identification of the FliE protein as a component of the flagellar hook-basal body complex. J Bacteriol 174: 2298–2304.

Ogasawara, H. and A. Hasegawa, E. Kanda, T. Miki, K. Yamamoto and A. Ishihama. 2007a. Genomic SELEX search for target genes under the control of PhoQP-RstBA signal relay cascade. J Bacteriol 189: 4791–4799.

Ogasawara, H. and Ishida, Y., Yamada, K. Yamamoto and A. Ishihama. 2007b. PdhR (pyruvte dehydrogenase complex regulator) controls the respiratory electron transport system in *Escherichia coli*. J Bacteriol 189: 5534–5541.

Ogasawara, H. and K. Yamada, A. Kori, K. Yamamoto and A. Ishihama. 2010a. Regulation of the *Escherichia coli csgD* promoter: interplay between five transcription factors. Microbiology 156: 2470–2483.

Ogasawara, H. and K. Yamamoto and A. Ishihama. 2010b. Regulatory role of MlrA in transcription activation of *csgD*, the master regulator of biofilm formation in *Escherichia coli*. FEMS Microbiol Lett 312: 160–168.

Ogasawara, H. and K. Yamamoto and A. Ishihama. 2011. Role of the biofilm master regulator CsgD in cross-regulation between biofilm formation and flagellar synthesis. J Bacteriol 193: 2587–2597.

Pan-Hou, H. and M. Kiyono, H. Omura, T. Omura and G. Endo. 2002. Polyphosphate produced in recombinant *Escherichia coli* confers meacury resistance. FEMS Microbiol Lett 207: 159–164.

Pesavento, C. and G. Becker, N. Sommerfeldt, A. Possling, N. Tschowri, A. Mehlis and R. Hengge. 2008. Inverse regulatory coordination of motility and curli-mediated adhesion in *Escherichia coli*. Genes Dev 22: 2434–2446.

Pin, C. and J.Baranyi. 2008. Kinetics of single cells: Observation and modeling of a stochastic process. Appl Environ Microbiol 74: 2534–2536.

Pratt, L.A. and T.J. Silhavy. 1998. Crl stimulates RpoS activity during stationary phase. Mol Microbiol 29: 1225–1236.

Prigent-Combaret, C. and E. Brombacher, O. Vidal, A. Ambert, P. Lejeune, P. Landini and C. Dorel. 2001. Complex regulatory network controls initial adhesion and biofilm formation in *Escherichia coli* via regulation of the *csgD* gene. J Bacteriol 183: 7213–7223.

Riley, M. and T. Abe, M.B. Arnaoud, M.K.B. Berlyn, F.R. Blattner, R.R. Chaudhuri, J.D. Glasner, T. Horiuchi, I.M. Keseler, T. Kosuge, H. Mori, N.T. Perma, G.P. Plunkett, III, K.E. Rudd, M.H. Serres, G.H. Thomas, N.R. Thompson, D. Wishart and B.L. Wanner. 2006. *Escherichia coli* K-12: a cooperatively developed annotation snapshot-2005. Nucleic Acids Res 34: 1–9.

Roostalu, J. and A. Joers, H. Luidalepp, N. Kaldalu and T. Tenson. 2008. Cell division in *Escherichia coli* cultures monitored at a single cell resolution. BMC Microbiol 8, 68.

Rosenfeld, N. and J.W. Young, U. Alon, P.S. Swain and M.B. Elowitz. 2005. Gene regulation at the single-cell level. Science 307: 1962–1965.

Roszak, D.B. and R.R. Colwell. 1987. Survival strategies of bacteria in the natural environment. Microbiol Rev 51: 365–379.
Shimada, T. and H. Makinoshima, Y. Ogawa, T. Miki, M. Maeda and A. Ishihama. 2004. Classification and strength measurement of stationary-phase promoters by use of a newly developed promoter cloning vector. J Bacteriol 186: 7112–7122.
Shimada, T. and N. Fujita, M. Maeda and A. Ishihama. 2005. Systematic search for the Cra-binding promoters using genomic SELEX system. Genes Cells 10: 907–918.
Shimada, T. and K.Hirao, A. Kori, K. Yamamoto and A. Ishihama. 2007. RutR is the uracil/thymine-sensing master regulator of a set of genes for synthesis and degradation of pyrimdines. Mol Microbiol 66: 744–757.
Shimada, T. and A. Ishihama, S.J.W. Busby and D.C. Grainger. 2008. The Escherichia coli RutR transcription factor binds at targets within genes as well as intergenic regions. Nucleic Acids Res 36: 3950–3955.
Shimada, T. and N. Fujita, K. Yamamoto and A. Ishihama. 2011a. Novel roles of cAMP receptor protein (CRP) in regulation of transport and metabolism of carbon sources. PLoS ONE (in press).
Shimada, T. and K. Yamamoto and A. Ishihama. 2011b. Novel members of the Cra regulon involved in carbon metabolism in *Escherichia coli.* J Bacteriol 193: 649–659.
Simm, R. and M. Morr, A. Kader, M. Nimtz and U. Romling. 2004. GGDEF and EAL domains inversely regulate cyclic d-GMP levels and transition from sessility to motility. Mol Microbiol 53: 1123–1134.
Simons, R.W. and F. Houman and N. Kleckner. 1987. Improved single and multicopy *lac*-based cloning vectors for protein and operon fusions. Gene 53: 85–96.
Stafford, G.P. and T. Ogi and C. Hughes. 2005. Binding and transcriptional activation of non-flagellar genes by the *Escherichia coli* flagellar master regulator FlhD2C2. Microbiology 151: 1779–1788.
Teramoto, J. and Y. Yamanishi, E-S.H. Magdy, A. Hasegawa, A. Kori, M. Nakajima, F. Arai, T. Fukuda and A.Ishihama. 2010a. Single live-bacterial cell assay of promoter activity and regulation. Genes Cells 15: 1111–1122.
Teramoto, J. and S. Yoshimura, K. Takeyasu and A. Ishihama. 2010b. A novel nucleoid protein of *Escherichia coli* induced under anaerobiotic growth conditions. Nucleic Acids Res 38: 3605–3618.
Umehara, S. and I. Inoue, Y. Wakamoto and K. Yasuda. 2007. Origin of individuality of two daughter cells during the division process examined by the simultaneous measurement of growth and swimming property using an on-chip single-cell cultivation system. Biophys J 93: 1061–1067.
Wolfe, A.J. and K.L. Visick. 2008. Get the message out: Cyclic-Di-GMP regulates multiple level of flagellum-based motility. J Bacteriol 190: 463–475.
Yamada, M. and H. Izu, T. Nitta, K. Kurihara and T. Sakurai. 1998. High-temperature, nonradioactive primer extension assay for determination of a transcription initiation site. Bio Techniques 25: 72–75.
Yamamoto, K. and A. Ishihama. 2003. Two different modes of transcription repression of the *Escherichia coli* acetate operon by IclR. Mol Microbiol 47: 183–194.

Yamanishi, Y. and J. Teramoto, Y. Magariyama, A. Ishihama, R. Fukuda and T. Arai. 2009. On-chip cell immobilization and monitoring system using thermosensitive gel contolled by suspended polymeric macrobridge. IEEE Transact. NanoBiosci 8: 312–317.

Yoshida, M. and K. Kashiwagi, G. Kawai, A. Ishihama and K. Igarashi. 2002. Polyamines enhance synthesis of the RNA polymerase sigma 38 subunit by suppression of an amber termination codon in the open reading frame. J Biol Chem 277: 37139–37146.

Zakikhany, K. and C.R. Harrington, M. Nimtz, N. Hinton and U. Romling. 2010. Unphosphorylated CsgD controls biofilm formation in *Salmonella enterica* serovar Typhimurium. Mol Microbiol 77: 771–786.

Zogaj, X. and N. Nimtz, M. Rohde, W. Bokranz and U. Romling. 2001. The multicellular morphotypes of *Salmonella typhimurium* and *Escherichia coli* produce cellulose as second component of the extracellular matrix. Mol Microbiol 39: 1452–1463.

Index

H

I

M

N

O

P

Q

S

T

W

Color Plate Section

Chapter 2

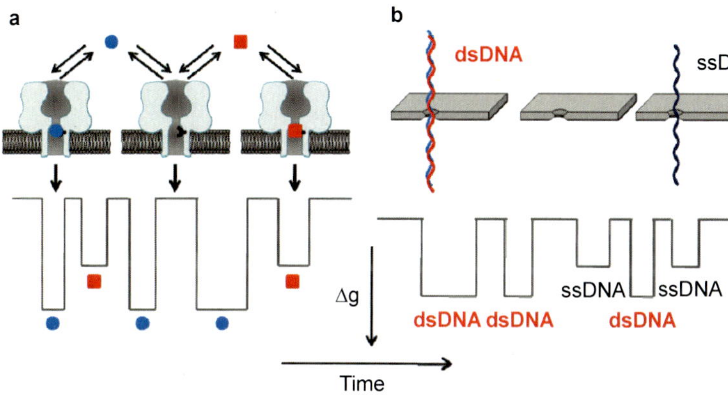

Figure 2.1: Cross-sectional view of single-molecule detection in protein nanopore and synthetic nanopore. (a) Different targets (shown as blue ball and red rectangle) bind to the engineered protein pore competitively and reversibly. Each target produces its own electrical signature with a specific current blocking amplitude and duration. (b) Synthetic nanopore with a double-stranded and single-stranded DNA threading through. This process produces different current blocking profile according to the size of threading molecules. Through single-molecule detection in the nanopore, targets can be identified from their electrical signature and quantified from the binding frequency.

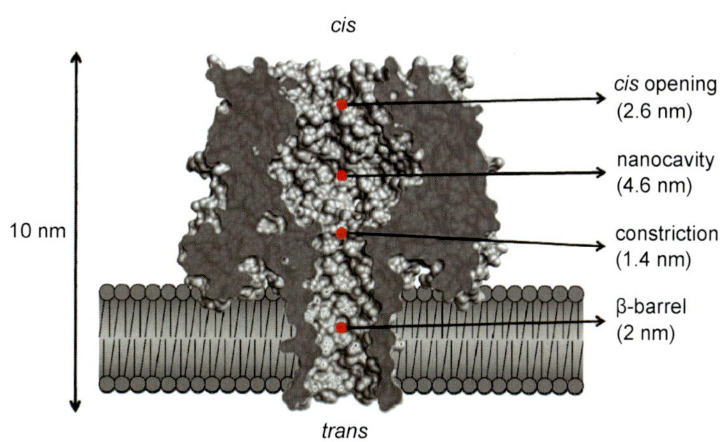

Figure 2.2: Molecular diagram of the α-hemolysin heptameric protein pore in the lipid bilayer (Song et al. 1996). The dimensions at specific sites are provided.

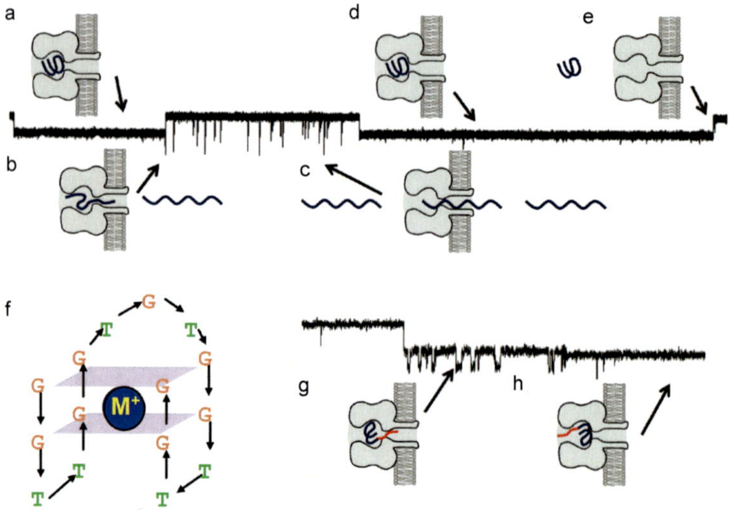

Figure 2.3: Investigation of folding and unfolding of a single G-quadruplex through a nanocavity enclosed in α-hemolysin. (a) The long-lived shallow current blocking indicates capturing a folded G-qaudruplex in the nanocavity. (b) G-quadruplex unfolds in the nanocavity and leaves the protein pore to the other side. This translocation event of unfolded aptamer is identified by terminal-spike. (c) Unfolded aptamers in linear form translocate through the protein pore. (d) Another folded G-quadruplex enters into the nanocavity. (e.) G-quadruplex escapes from the nanocavity to the entrance. (f) Schematic graph of folded Thrombin-Binding Aptamer G-quadruplex. It has two G-tetrads formed by 1, 6, 10, and 15 for top tetrad, and 2, 5, 11, and 14 for bottom one. A monovalent cation intervenes between two tetrads. (g) Current characteristic of tagged TBA. It produces an additional current blocking which indicates that the G-quadruplex is vibrating in the nanocavity. (h) The disappeared additional current blocking is shown by tag-TBA heading to the other side, which indicates that the G-quadruplex rotates in the nanocavity as well.

(Note: The data traces have been modified from its original. Spike-noise is hidden by figure a and b. Data trace for g and h are obtained in different time region of the same data trace but merged to show two events together.)

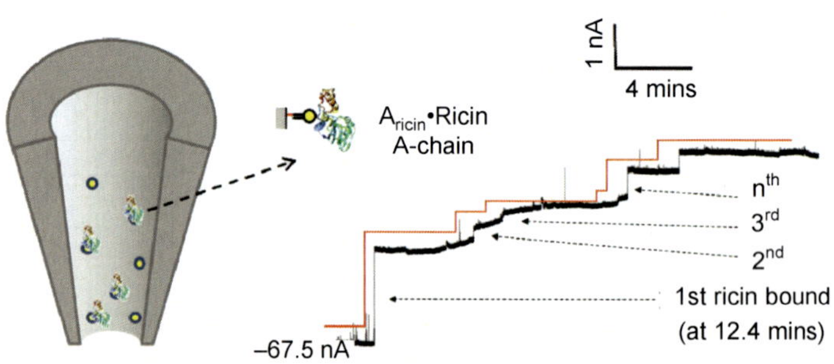

Figure 2.6: Detection of bioterrorist agent ricin with an aptamer-encoded glass-terminal nanopore. Ricin A-chain aptamers were immobilized on the internal surface of the nanopore. The aptamer-encoded nanopore was immersed in the solution containing 100 nM concentration of ricin A-chain protein and recorded at -100 mV. The stairway-like current blocking occurred as a series of molecules of ricin A-chain protein bind to the aptamer individually.

Chapter 3

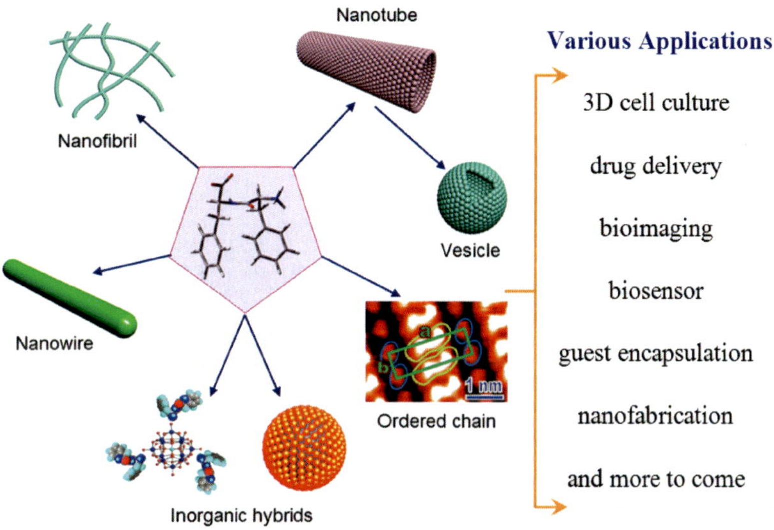

Figure 3.1: Representation of the various structures formed by self-assembly of diphenylalanine building blocks and their potential applications in nanotechnology. (Yan et al.2010. Self-assembly and application of diphenylalanine-based nanostructures. Chem Soc Rev 2010, 39, 1877–1890. Reproduced by permission of The Royal Society of Chemistry).

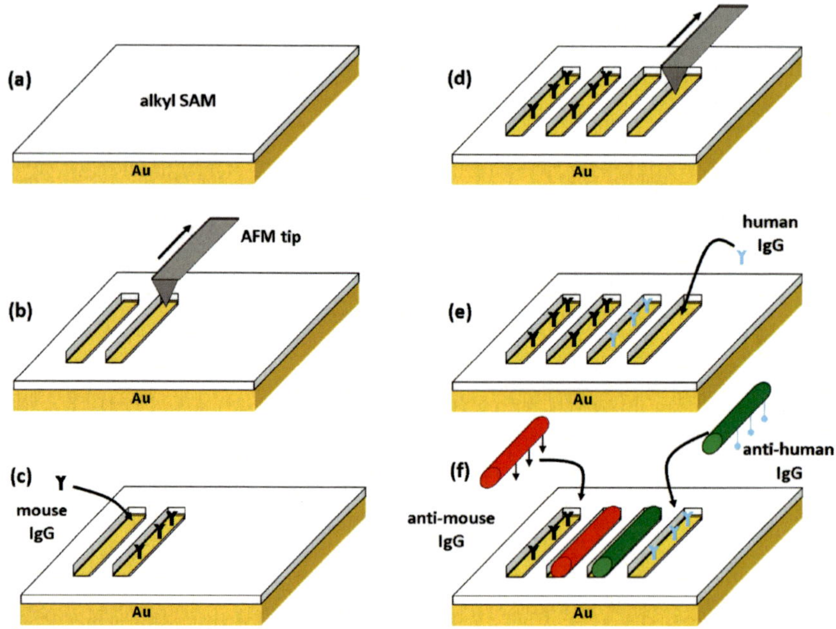

Figure 3.4: A schematic diagram to assemble anti-mouse IgG-coated nanotubes and anti-human IgG-coated nanotubes onto their antigen-patterned substrates via biological recognition. (a) Self-assembly of alkylthiol monolayers on Au substrates. (b) Shaving trenches on the alkylthiol SAM with an AFM tip. (c) Deposition of mouse IgG in the shaved trenches. (d) Shaving another array of trenches on the alkylthiol SAM using an AFM tip. (e) Deposition of human IgG in the shaved trenches. (f) Location-specific immobilization of Alexa Fluor 546-labeled anti-mouse IgG nanotubes into the mouse IgG trenches and FITC-labelled anti-human IgG nanotubes into the human IgG trenches via their biological recognition. Reprinted with permission from Zhao, Z. et al. 2005. J Am Chem Soc 127, 8930–8931. Copyright 2005 American Chemical Society.

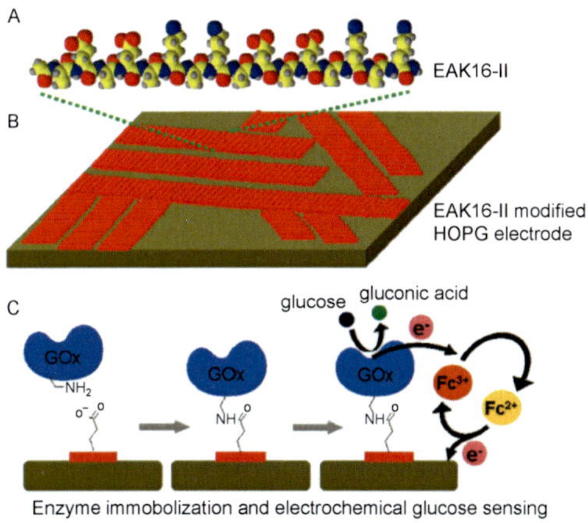

Figure 3.5: A schematic diagram of an ionic-complementary peptide modified HOPG electrode for GOx immobilization and glucose sensing. (A) EAK16-IImolecular structure, (B) EAK16-II nanofiber coating on the HOPG electrode, and (C) GOx immobilization and the glucose sensing mechanism. Reprinted with permission from Yang et al. 2009. Langmuir, 25, 7773–7777. Copyright 2009 American Chemical Society.

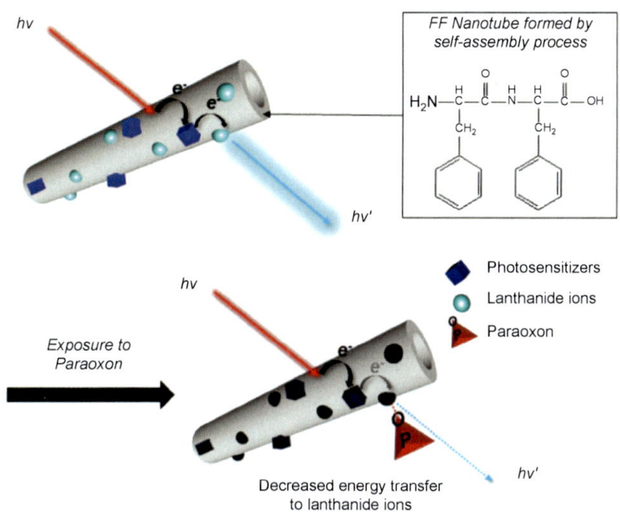

Figure 3.8: Suggested mechanism for the selective detection of paraoxon using photoluminescent FF nanotubes. FF nanotubes that are hybridized with lanthanide complexes exhibit an intense PL due to the transfer of cascaded energy from FF to lanthanide ions via photosensitizers. Upon the exposure to paraoxon, their PL could be readily quenched due to the inhibition of the cascaded-energy transfer. (Kim et al. 2011. Selective detection of neurotoxin by photoluminescent peptide nanotubes. Small. 6. 718–722. Copyright Wiley-VCH Verlag GmbH & Co. KGaA. Reproduced with permission.)

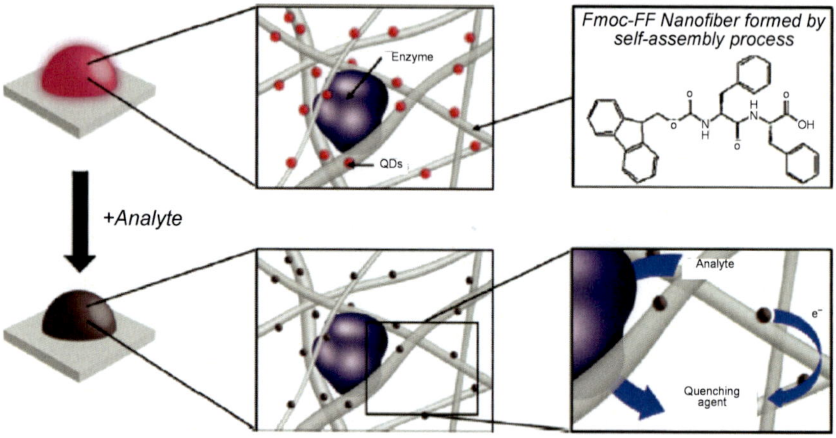

Figure 3.9: An illustrative description of the development of a photoluminescent peptide hydrogel through the self-assembly of Fmoc-FF building blocks and their PL quenching associated with the enzymatic detection of analytes. (Reprinted from Biosensors & Bioelectronics, 26, Kim et al. 2011. Self-assembled, photoluminiscent peptide hydrogel as a versatile platform for enzyme-based optical biosensors, 1860–1865, Copyright 2011, with permission from Elsevier.)

Chapter 5

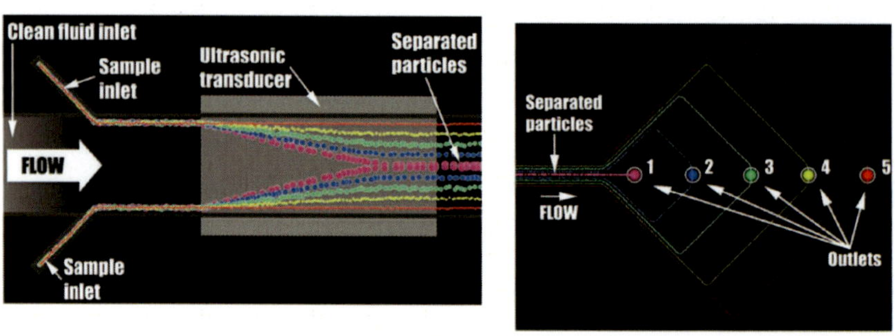

Figure 5.1: Acoustophoresis. (a) Particles entering main channel from two side inlets. Particles are positioned near channel walls because clean sheath fluid is introduced at a third inlet. The flow of particles is controlled by the acoustic waves, which are introduced by an ultrasonic transducer. After this point, the particles distribute proportional to size. (b) Flow splitters are used for separation of different sized particles. Nine fractions of the flow can be gathered at five outlets (Adapted from Peterson et al. 2007).

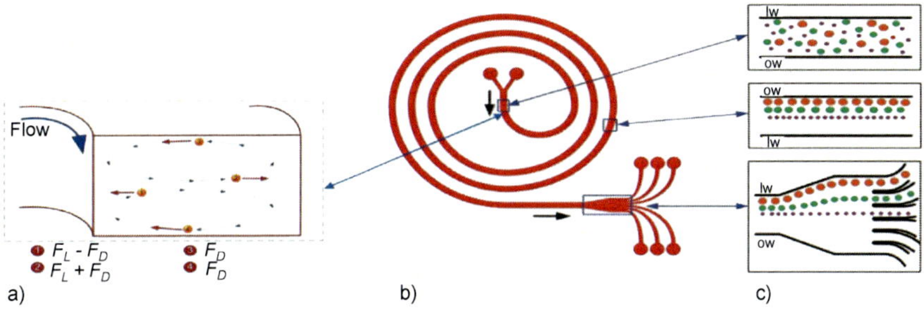

Figure 5.5: Spiral microchannel. (a) Neutral buoyant particles suspended in a medium in a spiral shaped channel experience forces and drag. Resultantly, particles redistribute within the microchannel. (b) Schematic representation of spiral channel for particle separation. (c) Different sized particles equilibrate at different positions in microchannel, and are collected at different outlets. (Adapted from Kuntaegowdanahalli et al. 2009 and Bhagat et al. 2008).

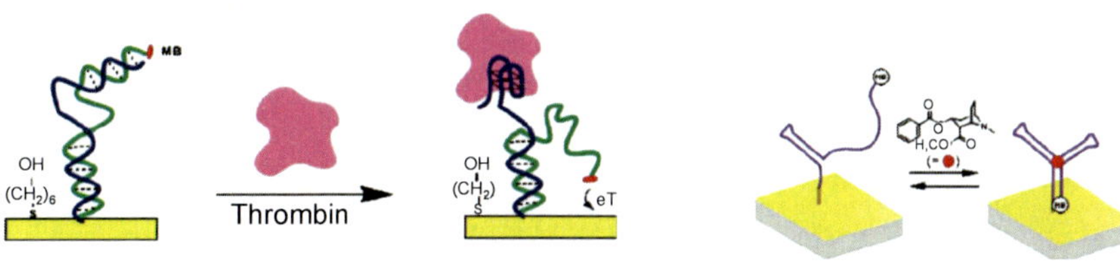

Figure 5.10: Aptamers with a methylene blue redox tag for thrombin (a) and cocaine detection (b). The binding event induces a conformational change in the aptamer and brings the redox active tag closer to the gold surface Baker et al. (2006); Lai et al. (2006). An analogue sensor was described earlier in Section 3.2. Xie et al. (2009) used carboxylic acid modified PEDOT nanowires instead of SWNTs as FET.

Chapter 6

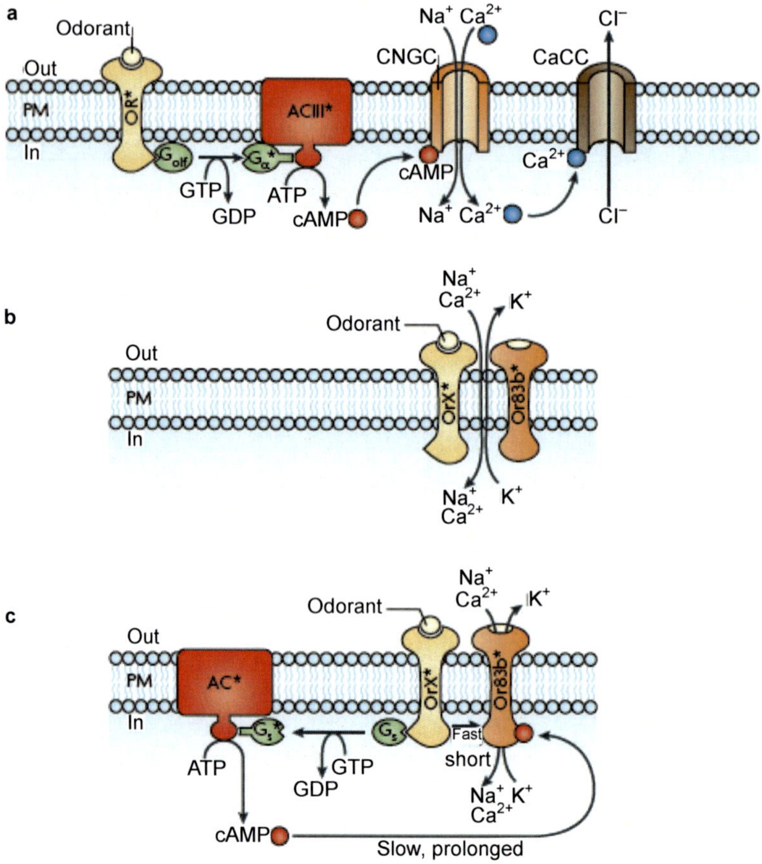

Figure 6.1: Adapted from Kaupp 2010. Nature Reviews. a) Odorant detection in vertebrates. When the odorant binds the olfactory receptor (OR) activates the trimeric transmembrane protein, olfaction-specific G protein (Golf), adenylyl cyclase type III (ACIII), the olfactory cyclic nucleotide-gated channel and a Ca2+-activated Cl - channel (CaCC). b) Insect olfactory model suggests that the odorant receptor forms an ion channel that is opened directly in response to the binding of odorants. c) On the other hand this complex insect model suggests that there are two pathways by which odour-induced depolarization can be generated. Upon odorant binding, activity is transferred to the Or83b subunit either by a direct (fast and short) or indirect (slow and prolonged) pathway.

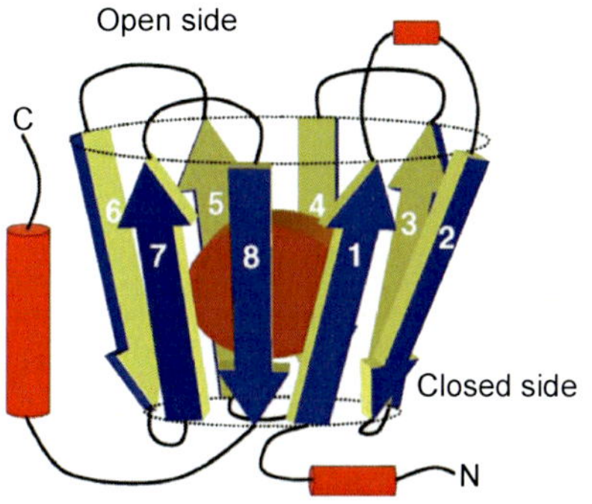

Figure 6.2: Schematic representation of an odorant-binding protein (Breer 2003).

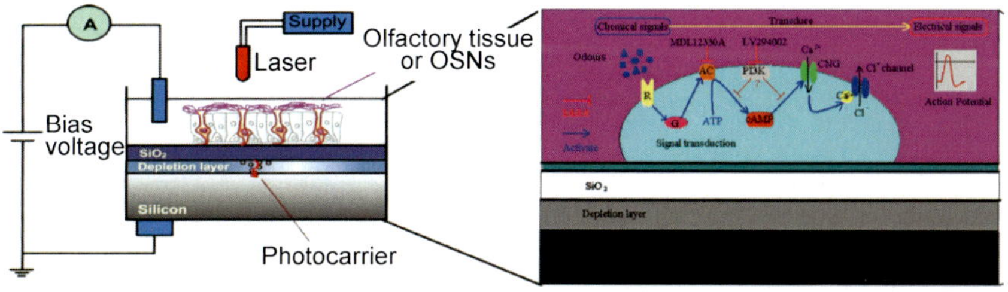

Figure 6.3: Biosensor *In vivo*. LAPS technique. Image adapted from the paper: Biosensors and Bioelectronics. Liu et al. (2010a).

Chapter 7

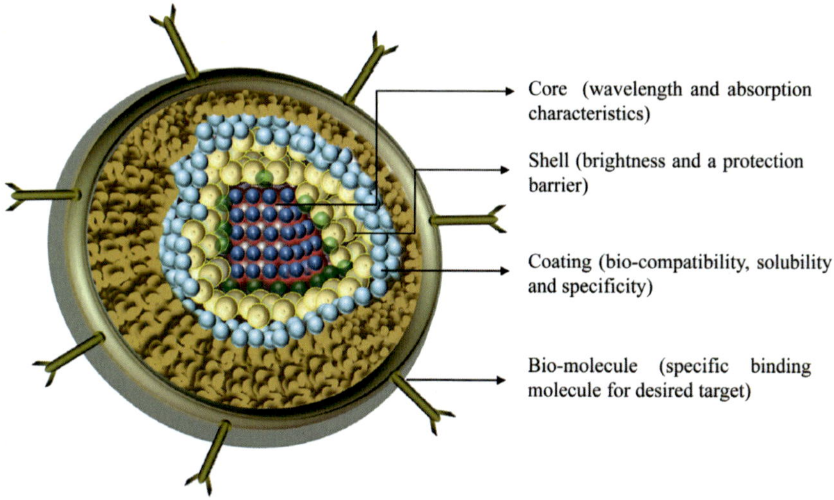

Figure 7.1: Structure of quantum dot.

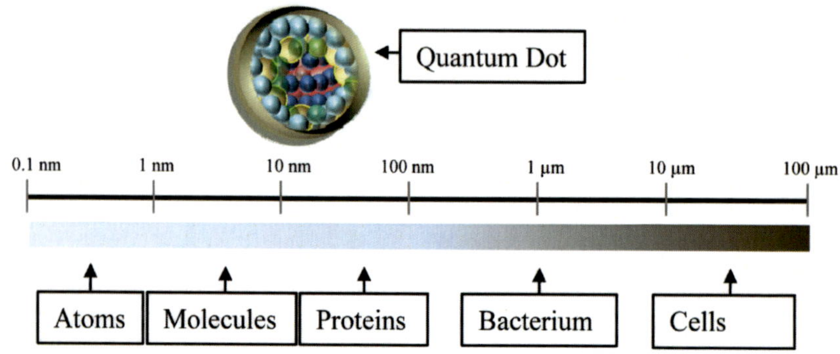

Figure 7.2: Size of quantum dot.

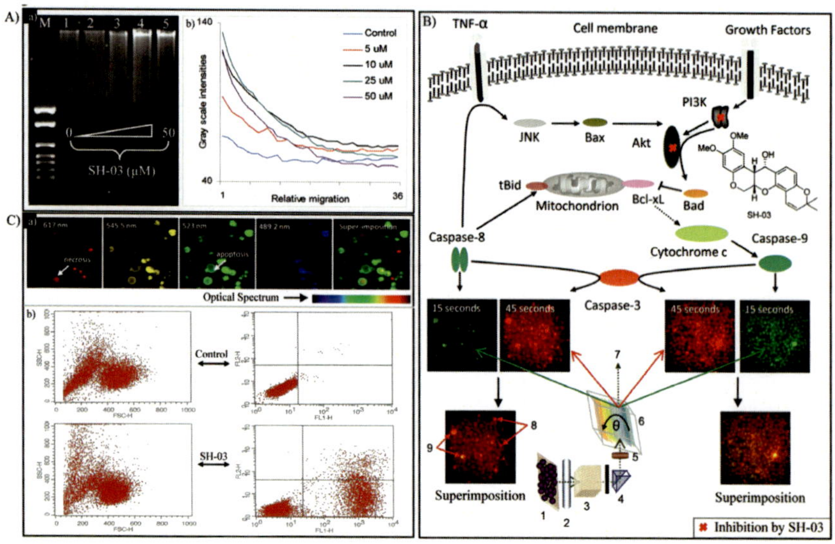

Figure 7.3: (A) DNA fragmentation in HL-60 cells after SH-03 treatment. Cells were treated with different concentrations of SH-03 and incubated for 6 hr at 37°C. (a) DNA was isolated from the cells and electrophorized on 0.8% agarose gel. M signifies the marker lane (1 kb DNA ladder), and lanes 1–5 denotes results obtained using 0, 5, 10, 25, and 50 μM SH-03, respectively. (b) Line graph depicting the average gray scale intensities of each DNA sample. (B) Structure of SH-03 [(7S,7a*R*,13a*S*)-9,10-dimethoxy-3,3-dimethyl-7,7a,13,13a-tetrahydro-3*H*-chromeno [3,4-b] pyrano[2,3-h]chromen-7-ol] and the activation of caspase-9 and caspase-3 after SH-03 treatment. Hyperspectral fluorescent images (SH-03; 10 μM) obtained using cellular imaging cytometry. In it, 1. Sample, 2. Objective lens (60 X), 3. Beam splitter, 4. Prism, 5. Focusing lens, 6. AOTF, 7. Un-diffracted beam, 8. Caspase-3 activation, 9. Caspase-3 and caspase-9 activation. θ. Bragg angle (the angle between diffracted and un-diffracted beam). (C) Representative hyperspectral fluorescent images of SH-03 (10 μM, 6 hr) induced apoptosis in HL-60 cells (a). Images were obtained over a spectral region from 463–688 nm with 60s of total time interval. Images taken at 523 nm and 617 nm were super-imposed. (b) Flow cytometric analysis illustrating the scatter and dot plots of HL-60 cells after 6 hr of SH-03 treatment. The lower right quadrant represent annexin V-FITC positive cells, whereas upper left quadrant represents propidium iodide (PI)-positive dead cells (necrosis).

Chapter 8

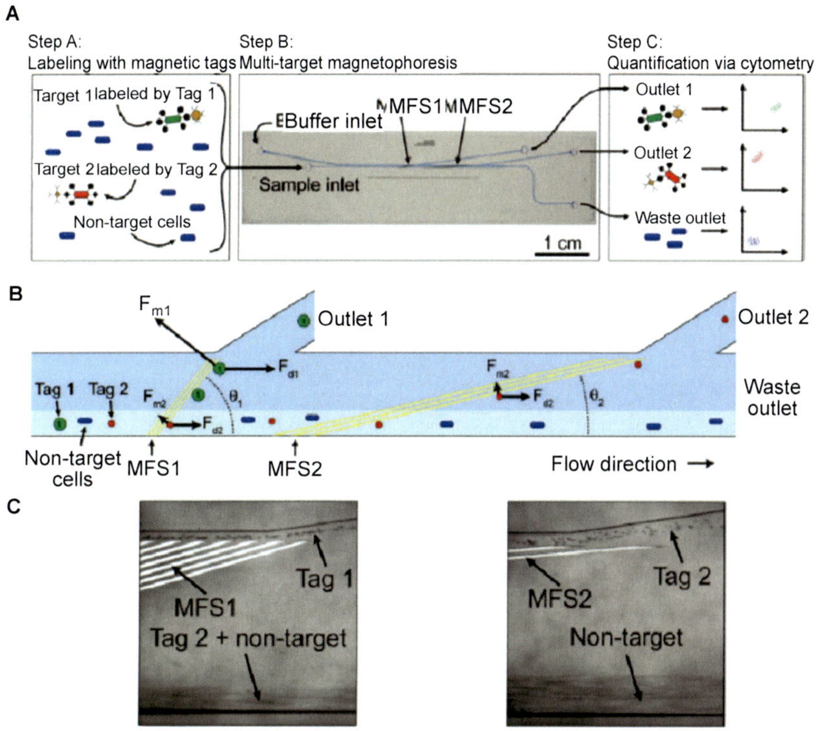

Figure 8.2: Schematic of the MT-MACS separation architecture.

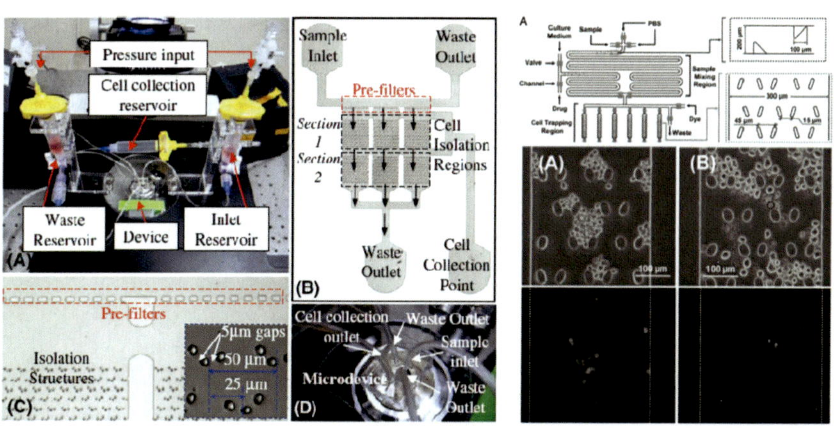

Figure 8.6: Microdevice with multiple arrays of crescent-shaped isolation wells integrated in the microchannel; microfluidic device with microdam structures.

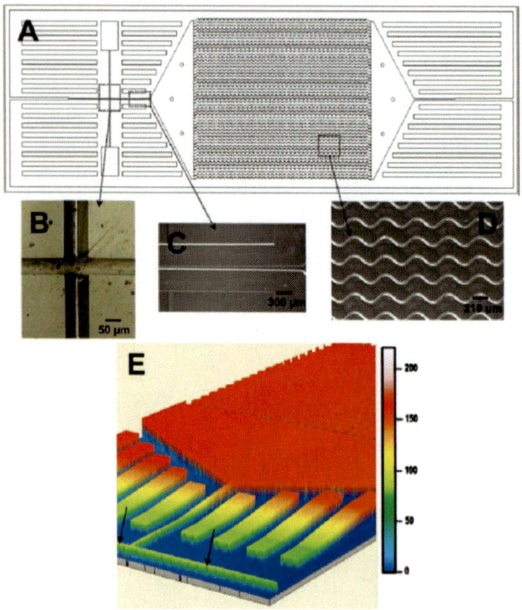

Figure 8.7: Modified PMMA microfluidic with a series of high-aspect ration microchannels in it was used to isolate and enrich CTCs from whole blood.

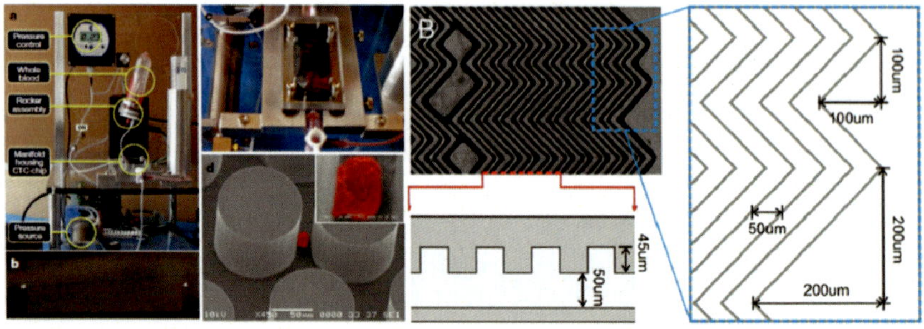

Figure 8.8: The 'CTC-chip' and the 'HB-chip'.

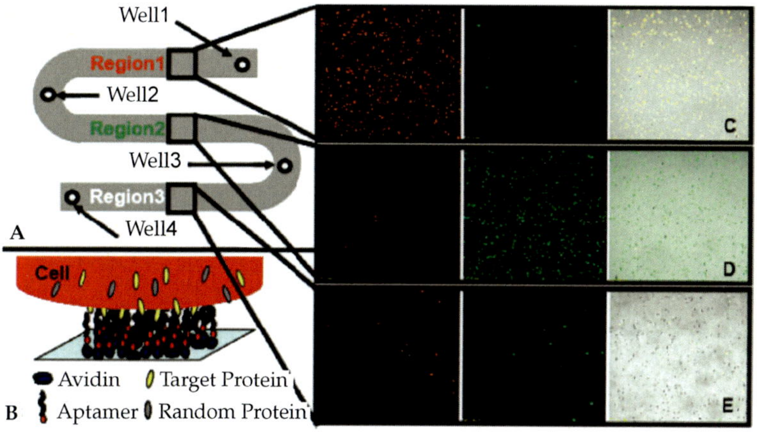

Figure 8.9: Aptamer-Based Microfluidic Device for Enrichment, Sorting, and Detection of Multiple Cancer Cells.

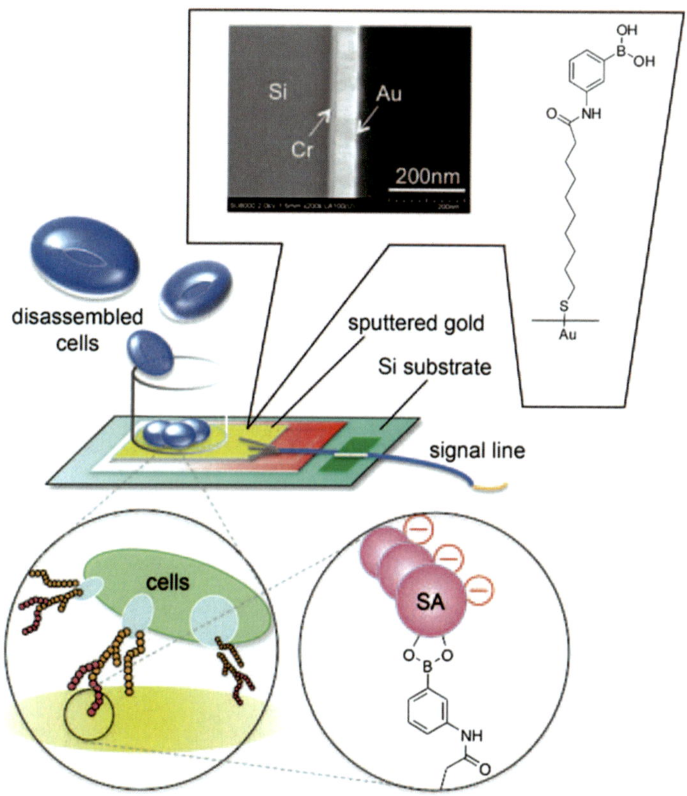

Figure 8.11: Sialic acid based electrochemical detection of tumor cells.

Chapter 9

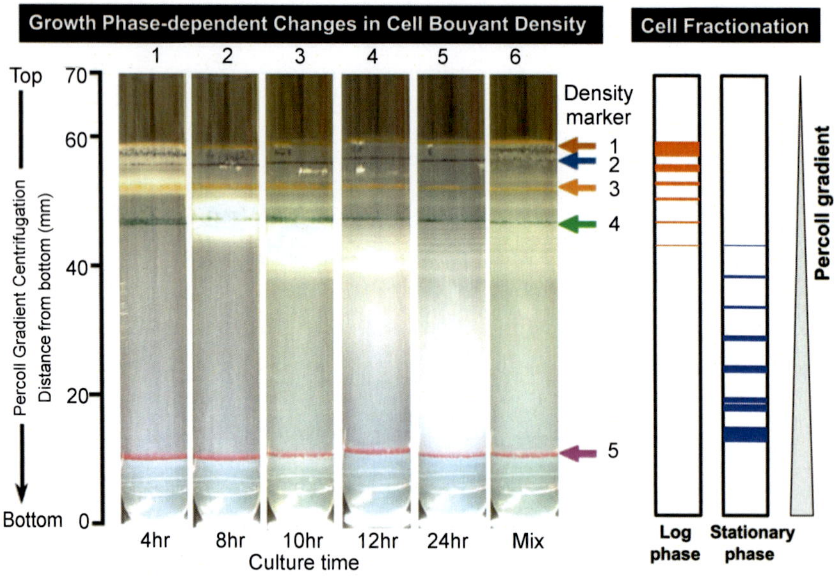

Figure 9.4: Bacterial cell fractionation through Percoll gradient centrifugation. Random culture of *E. coli* can be fractionated by Percoll density gradient centrifugation (Makinoshima et al. 2002). Upon entry into the stationary growth phase, the cell buoyant density increases step-wise, thereby allowing the isolation of homogeneous cell population. In each step of the density increase, a different gene(s) is involved.